Module 3

Linear Equations and Lines

Second Edition

Steps in Mathematics Modules

Module 3

Linear Equations and Lines

Second Edition

Leon J. Ablon **Sherry Blackman** **Helen B. Siner**
The College of Staten Island, City University of New York

Anthony Giangrasso
La Guardia Community College, City University of New York

The Benjamin/Cummings Publishing Company, Inc.
Menlo Park, California • Reading, Massachusetts
London • Amsterdam • Don Mills, Ontario • Sydney

Sponsoring Editor: Susan A. Newman
Production Editor: Madeleine Dreyfack
Book Designer: Madeleine Dreyfack
Cover Designer: Judith Sager

TO

Fannie and Meyer Ablon
Shelly Blackman
Tony and Jane Giangrasso
Anne Alexander Bean

ISBN 0-8053-0133-X
CDEFGHIJK-DO-8987654

Library of Congress Cataloging in Publication Data
Main entry under title:

Linear equations and lines.

(Series in mathematics modules; module 3)
1. Equations. 2. Algebra—Graphic methods.
I. Ablon, Leon J. II. Series.
QA39.2.S47 no. 3, 1980 [QA211] 512.9s [512.9'4]
80-24967

The Benjamin/Cummings Publishing Company, Inc.
2727 Sand Hill Road
Menlo Park, California 94025

Preface

Purpose of the Steps in Mathematics Modules

This book is one of the Steps in Mathematics Modules. The purpose of this series is to demystify mathematics and provide a foundation for the study of college algebra. Throughout our years of teaching we have become sensitive to the difficulties our students encounter in their mathematics courses. These modules incorporate the techniques we have evolved to overcome these problems.

Features

Development of Concepts: There are very few "rules" in these texts. Whenever possible, we encourage students to figure out what to do from the meaning of the symbols.

Writing Style: We have been as clear as possible in presenting each mathematical concept. We have found that students relate well to our use of everyday language. Our guiding principle has been to use language and syntax no more complex than is necessary to convey each concept. It is the kind of language that is used in the classroom, but is rarely found in mathematics textbooks.

Organization: Each module is divided into eight lessons. Each lesson is designed to be covered in one class period.

Examples and Exercises: We have used many "worked out" examples in each lesson to demonstrate the mathematics. Each lesson ends with two sets of *graded* exercises. The first set is followed immediately by handwritten solutions, so the student can see what a typical solution might look like. The second set, or "additional" exercise set, does not have solutions in the module. The answers to these exercises can be found in the Rationale, a supplement to these modules.

How to Use the Modules

Pacing: Each module is made up of eight lessons, and contains material sufficient for a 40 to 50-minute class period. The background of some students may, of course, dictate a slower or faster pace. On the average we have found that each module can be completed in 10 to 12 class meetings, including time for review and examination.

Planning a Syllabus for a Lecture Mode: The chart below may assist you in setting up a program for a term. It shows the number of modules that can be covered in 10, 12 and 14-week terms with classes that meet 3, 4 and 5 times a week.

Weeks per term	Meetings per week		
	3	4	5
10	3 modules	3 or 4 modules	4 or 5 modules
12	3 modules	4 modules	5 modules
14	4 modules	5 modules	6 or 7 modules

Self-Study or Lab Mode of Instruction: The modules can be used in math lab or for independent study. The preceding chart serves as a guide for students to gauge their progress.

Flexibility: The modular presentation of material allows for flexibility in teaching. In particular, the concept of changing the form of expressions (Modules 2 and 4) is separated from the concept of solving equations (Modules 3 and 5).

A chart of the modules interdependence can be found in the Rationale. We recommend that each module be studied from cover to cover. This gives students a feeling of completeness and insures that they are prepared for succeeding modules.

The Diagnostic Test included in the Rationale is designed to assist in determining an appropriate starting module for each student.

Improvements in This Edition

The first edition of these modules has been used successfully by us and many other math teachers in classrooms and labs for over eight years. As a result of our experiences and helpful feedback from other teachers, we have made several changes and refinements in this second edition.

Decimals: This new edition integrates a review of decimals into both the lessons and homework exercises.

Word Problems: Word problems have been carefully integrated throughout the modules in the examples and exercises.

Exercises: We have increased the number of homework exercises by 20%.

Geometry: The series now includes a review of the concepts of perimeter, area, formulas and the Pythagorean theorem.

Design and Appearance: The modules are typeset, a pleasant change from the camera-ready typewriter version of the first edition. The use of italic and boldface type enables the student to quickly pick out examples, important concepts and key terms.

The Steps in Mathematics Modules Series

The first five steps in mathematics are the equivalent of an elementary algebra course: They are:

Module 1—Operations on Numbers

Module 2—Operations on Polynomials

Module 3—Linear Equations and Lines

Module 4—Factoring and Operations on Algebraic Fractions

Module 5—Quadratic Equations and Curves

Program Rationale and Tests—an instructor's supplement which contains a brief description of the content of each lesson, answers to the Additional Exercises, diagnostic placement examinations and three mastery tests for each of the five modules.

Additional modules in the **Steps in Mathematics Series** cover topics in intermediate algebra and other specialized topics.

Module 1a—Practical Mathematics

Module 2a—Practical Mathematics

Module 6 —Basic Trigonometry, second edition

Module 7 —Trigonometry with Applications

Module 8 —Exponents and Logarithms

Module 9 —Advanced Algebraic Techniques

Module 10—Functions and Word Problems

Module 11—Graphing Functions

Module M —Medical Dosage Calculations, second edition

Module SI—Metric System

Acknowledgements

We wish to thank all the people who contributed to the revision of Modules 1–5 with special thanks to our reviewers:

Dr. George Bergman, University of California, Berkeley
Dr. Una Bray, Marymount Manhattan College
Professor Ruth Dorsett, Atlanta Junior College
Dr. Susan Lawrence, New York University
Dr. Betty Philips, Michigan State University
Professor Michael Shaughnessy, Oregon State University
Professor Billie J. Stacey, Sinclair Community College

Our appreciation, also, to Susan Newman, sponsoring editor, and Madeleine Dreyfack, production editor and designer. The special efforts they made to translate our ideas into a finished work are greatly appreciated.

Leon J. Ablon
Sherry Blackman
Anthony P. Giangrasso
Helen B. Siner
October, 1980

To the Student

Read this text. It was written with the help of our students so students could read it. Our students tell us that the best way to learn this material is:

1. Read the lesson.
2. Work out each EXAMPLE yourself.
3. Try the first set of EXERCISES at the end of the lesson and compare your answers with ours.
4. Do the ADDITIONAL EXERCISES for more practice.

We wish you success, and we would like to hear your reactions to the text and your suggestions for future editions.

Contents

Lesson 1

Linear Equations in One Variable

An **equation** is a sentence that says two things are equal. The sentence "$2 + 7 = 9$" is an equation and it is true. The sentence "$3 + 7 = 9$" is an equation and it is false. The sentence "$x + 7 = 9$" is also an equation, but it is neither true nor false. If 2 is put in place of x, the equation is true. If 3 is put in place of x, the equation is false.

Finding the numbers that make an equation true is called **solving** the equation. The numbers that make the equation true are called the **solutions** of the equation. The letter x in the equation $x + 7 = 9$ is called a **variable**. Any letter or other symbol can be used as a variable but we will start out only using x's.

5 makes	$x + 3 = 8$	true, so 5 is a solution of	$x + 3 = 8$
7 makes	$x - 2 = 5$	true, so 7 is a solution of	$x - 2 = 5$
0 makes	$x + 4 = 4$	true, so 0 is a solution of	$x + 4 = 4$
2 makes	$2.6 = 1.3x$	true, so 2 is a solution of	$2.6 = 1.3x$
3 makes	$\frac{6}{x} = 2$	true, so 3 is a solution of	$\frac{6}{x} = 2$
8 makes	$\frac{x}{2} = 4$	true, so 8 is a solution of	$\frac{x}{2} = 4$
7 makes	$x = 7$	true, so 7 is a solution of	$x = 7$

Sometimes it is not easy to see what makes an equation true, but there are things we can do that may help. The basic idea is, if two things are equal and you do the same thing to both of them, they are still equal. So, if you and your cousin are the same height and you both grow three inches, you are both still the same height.

EXAMPLE 1 Solve the equation $x + 3 = -9$.

If you can't see what makes $x + 3 = -9$ true, try adding -3 to both sides.

$$\begin{array}{lrcr} & x + 3 & = & -9 \\ \text{Add} & -3 & & -3 \\ \hline & x + 0 & = & -12 \end{array}$$

but $x + 0$ is x

so $x = -12$

Since -12 makes $x = -12$ true, -12 should also make $x + 3 = -9$ true. Let's try it.

$$x + 3 = -9$$

Put -12 in for x.

$$\begin{array}{lcl} -12 + 3 & = & -9 \\ -9 & = & -9 \quad \underline{\text{True!}} \end{array}$$

So the solution is -12.

EXAMPLE 2 Solve the equation $8 = 15 + x$.

If you can't see what makes $8 = 15 + x$ true, try adding -15 to both sides.

$$\begin{array}{lrcr} & 8 & = & 15 + x \\ \text{Add} & -15 & & -15 \\ \hline & -7 & = & 0 + x \end{array}$$

but $0 + x$ is x

so $-7 = x$

Since -7 makes $-7 = x$ true, -7 should also make $8 = 15 + x$ true. Let's try it.

$$8 = 15 + x$$

Put -7 in for x.

$$8 = 15 + (-7)$$
$$8 = 15 - 7$$
$$8 = 8 \qquad \underline{\text{True!}}$$

So the solution is -7.

In both examples we worked until we got to the easiest kind of equation to solve, the one that looks like "x = number," or "number = x." In example 1, we worked on $x + 3 = -9$ until we got $x = -12$. In example 2, we worked on $8 = 15 + x$ until we got $-7 = x$.

EXAMPLE 3 Solve the equation $x - 5 = 11$.

Try to get the easiest kind of equation to solve: "x = number" or "number = x."

We can get rid of the -5 on the left side by adding $+5$ (since $-5 + 5$ is 0), but we must do the same thing to <u>both</u> sides.

$$\begin{array}{lrcr} & x - 5 & = & 11 \\ \text{Add} & +5 & & +5 \\ \hline & x + 0 & = & 16 \end{array}$$

but $x + 0$ is x

so $x = 16$

Since 16 makes $x = 16$ true, 16 also makes $x - 5 = 11$ true. So the solution is 16. To check the solution, put 16 in for x in the original equation.

$$x - 5 = 11$$
$$16 - 5 = 11$$
$$11 = 11 \qquad \underline{\text{True!}}$$

EXAMPLE 4 Solve the equation $3x = 2$.

Let's head for "x = number."

Note: $3x$ means 3 times x.

If you can't see what makes $3x = 2$ true, try dividing both sides of the equation by 3 because $\frac{3x}{3}$ is $1x$ or x.

$$3x = 2$$

Divide by 3.

$$\frac{3x}{3} = \frac{2}{3}$$

$$\frac{\overset{1}{\cancel{3}}x}{\underset{1}{\cancel{3}}} = \frac{2}{3}$$

$$1x = \frac{2}{3}$$

So $x = \frac{2}{3}$

Since $\frac{2}{3}$ makes $x = \frac{2}{3}$ true, $\frac{2}{3}$ also makes $3x = 2$ true. So the solution is $\frac{2}{3}$. To check the solution, put $\frac{2}{3}$ in for x in the original equation.

$$3x = 2$$

$$3\left(\frac{2}{3}\right) = 2$$

$$\frac{\overset{1}{\cancel{3}}\cdot 2}{\underset{1}{\cancel{3}}} = 2$$

$$2 = 2$$ <u>True!</u>

EXAMPLE 5 Solve the equation $8x = -4$.

Let's head for "x = number."

Note: $8x$ means 8 times x.

We will divide both sides of the equation by 8 because $\frac{8x}{8}$ is $1x$ or x.

$$8x = -4$$

Divide by 8.

$$\frac{8x}{8} = \frac{-4}{8}$$

$$\frac{\overset{1}{\cancel{8}}x}{\underset{1}{\cancel{8}}} = -\frac{\overset{1}{\cancel{4}}}{\underset{1}{\cancel{4}}\cdot 2}$$

$$1x = -\frac{1}{2}$$

So $\quad x = -\frac{1}{2}$

Since $-\frac{1}{2}$ makes $x = -\frac{1}{2}$ true, $-\frac{1}{2}$ also makes $8x = -4$ true. So the solution is $-\frac{1}{2}$. To check the solution, put $-\frac{1}{2}$ in for x in the original equation.

$$8x = -4$$

$$8\left(-\frac{1}{2}\right) = -4$$

$$-4 = -4 \quad \underline{\text{True!}}$$

EXAMPLE 6 Solve the equation $5x = 0$.

Let's head for "x = number."

Note: $5x$ means 5 times x.

We will divide both sides of the equation by 5 because $\frac{5x}{5}$ is $1x$ or x.

$$5x = 0$$

Divide by 5.

$$\frac{5x}{5} = \frac{0}{5}$$

$$\frac{\overset{1}{\cancel{5}}x}{\underset{1}{\cancel{5}}} = 0$$

$$1x = 0$$

So $x = 0$

Since 0 makes $x = 0$ true, 0 also makes $5x = 0$ true. So the solution is 0. To check the solution, put 0 in for x in the original equation.

$$5x = 0$$
$$5(0) = 0$$
$$0 = 0 \quad \underline{\text{True!}}$$

EXAMPLE 7 Solve the equation $-x = 15$.

Let's head for "x = number."

Note: $-x$ means -1 times x.

$$-x = 15$$
$$-1x = 15$$

We will divide both sides of the equation by -1 because $\frac{-1x}{-1}$ is $1x$ or x.

$$-1x = 15$$

Divide by -1.

$$\frac{-1x}{-1} = \frac{15}{-1}$$

$$1x = -15$$

So $x = -15$

Since -15 makes $x = -15$ true, -15 also makes $-x = 15$ true. So the solution is -15. To check the solution, put -15 in for x in the original equation.

$$-x = 15$$
$$-(-15) = 15$$
$$15 = 15 \quad \underline{\text{True!}}$$

EXAMPLE 8 Solve the equation $-2 = \frac{1}{3}x$.

Let's head for "number $= x$."

Note: $\frac{1}{3}x$ means $\frac{1}{3}$ times x.

We will multiply both sides of the equation by 3 because $3 \cdot \frac{1}{3} \cdot x$ is $1x$ or x.

$$-2 = \frac{1}{3}x$$

Multiply both sides of the equation by 3.

$$3(-2) = \overset{1}{\cancel{3}}\left(\frac{1}{\underset{1}{\cancel{3}}}x\right)$$

$$-6 = 1x$$

So $-6 = x$

Since -6 makes $-6 = x$ true, -6 also makes $-2 = \frac{1}{3}x$ true. So the solution is -6. To check the solution put -6 in for x in the original equation.

$$-2 = \frac{1}{3}x$$
$$-2 = \frac{1}{3}(-6)$$
$$-2 = -2 \qquad \underline{\text{True!}}$$

EXAMPLE 9 Solve the equation $\frac{2x}{5} = 9.6$.

Let's head for "x = number." First we multiply both sides of the equation by 5.

$$\frac{2x}{5} = 9.6$$
$$\overset{1}{\cancel{5}}\left(\frac{2x}{\underset{1}{\cancel{5}}}\right) = 5(9.6)$$
$$2x = 48$$

Now we divide both sides of the equation by 2.

$$\frac{\overset{1}{\cancel{2}}x}{\underset{1}{\cancel{2}}} = \frac{48}{2}$$
$$1x = 24$$

So $x = 24$

Since 24 makes $x = 24$ true, 24 also makes $\frac{2x}{5} = 9.6$ true. So the solution is 24. To check the solution put 24 in for x in the original equation.

$$\frac{2x}{5} = 9.6$$
$$\frac{2(24)}{5} = 9.6$$
$$\frac{48}{5} = 9.6$$
$$9.6 = 9.6 \qquad \underline{\text{True!}}$$

EXAMPLE 10 Solve the equation $x = 0.08(250)$.

Let's head for "x = number." Since the equation is already in this form, all we have to do is simplify the right side of the equation.

$$x = 0.08(250)$$

$$x = 20$$

So the solution is 20.

EXAMPLE 11 Solve the equation $0.42x = 63$.

Let's head for "x = number." We divide both sides of the equation by 0.42.

$$0.42x = 63$$

$$\frac{\overset{1}{\cancel{0.42}}x}{\underset{1}{\cancel{0.42}}} = \frac{63}{0.42}$$

$$x = 150$$

So the solution is 150.

EXAMPLE 12 Solve the equation $150 = \dfrac{x}{100}(1250)$.

Since the only x is on the right side of the equation, we head for "number = x." We multiply both sides of the equation by 100.

$$150 = \frac{x}{100}(250)$$

$$100(150) = \overset{1}{\cancel{100}}\left(\frac{x}{\underset{1}{\cancel{100}}}\right)(1250)$$

$$15000 = 1250x$$

Divide by 1250.

$$\frac{15000}{1250} = \frac{\overset{1}{\cancel{1250}}x}{\underset{1}{\cancel{1250}}}$$

$$12 = x$$

So the solution is 12.

We can use equations to help solve problems that involve percents. Let's find the amount of sales tax on a purchase of $250 when sales tax is 8%.

We change the problem to a question:

What is 8% of $250?

We change the question to an equation by replacing the words with symbols.

Word		Symbol
What	→	x
is	→	=
%	→	(Divide by 100)
of	→	(times)

What is 8% of $250?

becomes $x = 0.08 \cdot 250$

Note: We do not use the symbol % in an equation. Since % means "divided by 100," we can write 8% as $\frac{8}{100}$ or 0.08.

In example 10, we solved the equation $x = 0.08(250)$ and found $x = 20$. In this example, "what" stands for the sales tax. Since x is 20, the sales tax is $20.

EXAMPLE 13 42% of the total number of students in a law school are women. If there are 63 women in the school, how many students are there in the school?

We change the problem to a statement:

42% of the students are women.

We change the statement to an equation by replacing the words with symbols.

	42%	of	the students	are	women
becomes	0.42	$\cdot$	x	$=$	63

In example 11, we solved the equation $0.42x = 63$, and found $x = 150$. In this example, x stands for the total number of students in the school. Since x is 150, there are 150 students in the school.

EXAMPLE 14 Helen Brody pays \$150 rent and her monthly income is \$1250. Her rent is what percent of her total income?

The question is:

Rent is what percent of income?

or \$150 is what percent of \$1250?

We change the question to an equation.

\$150	is	what percent	of	\$1250?
150	$=$	$\frac{x}{100}$	$\cdot$	1250

In example 12, we solved the equation $150 = \frac{x}{100}(1250)$ and found

$12 = x$. In this example, x stands for the percent of total income. Since x is 12, Brody's rent is 12% of her total income.

Exercises

Solve the following equations and check each solution.

1. $x + 4 = 8$
2. $x + 2 = 5$
3. $x + 5 = -3$
4. $x + 4 = -7$
5. $x - 2 = 6$
6. $x - 7 = 12$
7. $4 = x + 3$
8. $2 = x - 3$
9. $x - 7 = -4$
10. $18 + x = 17$
11. $6 = 6 + x$
12. $2x = 4$
13. $4x = 12$
14. $9 = 3x$
15. $20 = 5x$
16. $7x = -14$
17. $-12 = 2x$
18. $-6 = 5x$
19. $4x = 5$
20. $2x = 3$
21. $4x = 0$
22. $-2x = 8$
23. $-5x = 15$
24. $-3x = -9$
25. $-4x = -4$
26. $-x = 1$
27. $-x = -1$
28. $-x = 12$
29. $-x = -12$
30. $12 = -24x$
31. $\frac{1}{2}x = 5$
32. $\frac{1}{3}x = 6$
33. $\frac{1}{5}x = -2$
34. $\frac{1}{4}x = -4$

35. $-6 = \frac{1}{3}x$

36. $\frac{1}{10}x = 100$

37. $\frac{2}{3}x = 8$

38. $\frac{3}{4}x = 6$

39. $14 = \frac{7}{8}x$

40. $-\frac{2}{3} = \frac{2}{3}x$

41. $\frac{5}{6}x = -5$

42. $\frac{89}{90}x = 0$

43. $3x = 3.6$

44. $-2x = -2.4$

45. $\frac{3x}{4} = 5.7$

46. $\frac{2x}{3} = 4.2$

47. $x = 0.05(34)$

48. $x = 0.12(22)$

49. $0.12x = 75$

50. $0.5x = 12$

51. $120 = \frac{x}{100}(80)$

52. $140 = \frac{x}{100}(2800)$

Do the following word problems.

53. What is 5% of 34?

54. What is 12% of 22?

55. The sales tax is 8%. Find the tax on a suit that costs $45.

56. The sales tax is 8%. Find the tax on a blouse that costs $12.

57. 35% of the total number of students in a medical school are women. If there are 63 women in the school, how many students are there in the school?

58. 60% of the total number of students in a dental school are men. If there are 84 men in the school, how many students are there all together?

59. 12 is what percent of 30?

60. 42 is what percent of 60?

61. Joe Harris pays $180 monthly rent and his monthly income is $800. His rent is what percent of his monthly income?

62. Jeanette Gonzalez pays $210 rent each month and earns $750. What percent of her earnings does she spend for rent?

Answers to Exercises

(1.) $x + 4 = 8$
$\quad -4 \quad -4$
$x = 4$
solution is 4

(2.) $x + 2 = 5$
$\quad -2 \quad -2$
$x = 3$
solution is 3

(3.) $x + 5 = -3$
$\quad -5 \quad -5$
$x = -8$
solution is -8

(4.) $x + 4 = -7$
$\quad -4 \quad -4$
$x = -11$
solution is -11

(5.) $x - 2 = 6$
$\quad +2 \quad +2$
$x = 8$
solution is 8

(6.) $x - 7 = 12$
$\quad +7 \quad +7$
$x = 19$
solution is 19

(7.) $4 = x + 3$
$-3 \quad -3$
$1 = x$
solution is 1

(8.) $2 = x - 3$
$+3 \quad +3$
$5 = x$
solution is 5

(9.) $x - 7 = -4$
$\quad +7 \quad +7$
$x = +3$
solution is 3

(10.) $18 + x = 17$
$-18 \quad -18$
$x = -1$
solution is -1

(11.) $6 = 6 + x$
$-6 \quad -6$
$0 = x$
solution is 0

(12.) $\frac{2x}{2} = \frac{4}{2}$
$x = 2$
solution is 2

(13.) $\frac{4x}{4} = \frac{12}{4}$
$x = 3$
solution is 3

(14.) $\frac{9}{3} = \frac{3x}{3}$
$3 = x$
solution is 3

(15.) $\frac{20}{5} = \frac{5x}{5}$
$4 = x$
solution is 4

(16.) $\frac{7x}{7} = \frac{-14}{7}$
$x = -2$
solution is -2

(17.) $\frac{-12}{2} = \frac{2x}{2}$
$-6 = x$
solution is -6

(18.) $\frac{-6}{5} = \frac{5x}{5}$
$-\frac{6}{5} = x$
solution is $-\frac{6}{5}$

(19.) $\frac{4x}{4} = \frac{5}{4}$

$x = \frac{5}{4}$

solution is $\frac{5}{4}$

(20.) $\frac{2x}{2} = \frac{3}{2}$

$x = \frac{3}{2}$

solution is $\frac{3}{2}$

(21.) $\frac{4x}{4} = \frac{0}{4}$

$x = 0$

solution is 0

(22.) $\frac{-2x}{-2} = \frac{8}{-2}$

$x = -4$

solution is -4

(23.) $\frac{-5x}{-5} = \frac{15}{-5}$

$x = -3$

solution is -3

(24.) $\frac{-3x}{-3} = \frac{-9}{-3}$

$x = 3$

solution is 3

(25.) $\frac{-4x}{-4} = \frac{-4}{-4}$

$x = 1$

solution is 1

(26.) $\frac{-x}{-1} = \frac{1}{-1}$

$x = -1$

solution is -1

(27.) $\frac{-x}{-1} = \frac{-1}{-1}$

$x = 1$

solution is 1

(28.) $\frac{-x}{-1} = \frac{12}{-1}$

$x = -12$

solution is -12

(29.) $\frac{-x}{-1} = \frac{-12}{-1}$

$x = 12$

solution is 12

(30.) $\frac{12}{-24} = \frac{-24x}{-24}$

$-\frac{1}{2} = x$

solution is $-\frac{1}{2}$

(31.) $\frac{1}{2}x = 5$

$2\left(\frac{1}{2}x\right) = 2(5)$

$x = 10$

solution is 10

(32.) $\frac{1}{3}x = 6$

$3\left(\frac{1}{3}x\right) = 3(6)$

$x = 18$

(33.) $\frac{1}{5}x = -2$

$5\left(\frac{1}{5}x\right) = 5(-2)$

$x = -10$

solution is -10

(34.) $\frac{1}{4}x = -4$

$4(\frac{1}{4}x) = 4(-4)$

$x = -16$

solution is -16

(35.) $-6 = \frac{1}{3}x$

$3(-6) = (\frac{1}{3}x)3$

$-18 = x$

solution is -18

(36.) $\frac{1}{10}x = 100$

$10(\frac{1}{10}x) = 10(100)$

$x = 1000$

solution is 1000

(37.) $\frac{2}{3}x = 8$

$\frac{3}{2}(\frac{2}{3}x) = \frac{3}{2} \cdot 8$

$x = 3(4)$ or 12

solution is 12

(38.) $\frac{3}{4}x = 6$

$(\frac{4}{3}) \cdot \frac{3}{4}x = (\frac{4}{3}) \cdot 6$

$x = 8$

solution is 8

(39.) $14 = \frac{7}{8}x$

$(\frac{8}{7}) \cdot 14 = (\frac{8}{7}) \cdot \frac{7}{8}x$

$16 = x$

solution is 16

(40.) $-\frac{2}{3} = \frac{2}{3}x$

$(\frac{3}{2}) \cdot \frac{-2}{3} = (\frac{3}{2}) \frac{2}{3}x$

$-1 = x$

solution is -1

(41.) $\frac{5}{6}x = -5$

$(\frac{6}{5})\frac{5}{6}x = (\frac{6}{5})(-5)$

$x = -6$

solution is -6

(42.) $\frac{89}{90}x = 0$

$\left(\frac{90}{89}\right)\frac{89}{90}x = \left(\frac{90}{89}\right)(0)$

$x = 0$

solution is 0

(43.) $\frac{3x}{3} = \frac{3.6}{3}$

$x = 1.2$

solution is 1.2

(44.) $\frac{-2x}{-2} = \frac{-2.4}{-2}$

$x = 1.2$

solution is 1.2

(45.) $\frac{3x}{4} = 5.7$

$\left(\frac{4}{3}\right)\frac{3x}{4} = \left(\frac{4}{3}\right)(5.7)$

$x = 7.6$

solution is 7.6

(46.) $\frac{2x}{3} = 4.2$

$\left(\frac{3}{2}\right)\frac{2x}{3} = \left(\frac{3}{2}\right)(4.2)$

$x = 6.3$

solution is 6.3

(47.) $x = .05(34)$

$x = 1.7$

(48.) $x = .12(22)$

$x = 2.64$

solution is 2.64

(49.) $0.12x = 75$

$\frac{0.12x}{0.12} = \frac{75}{0.12}$

$x = 625$

solution is 625

(50.) $0.5x = 12$

$$\frac{0.5x}{0.5} = \frac{12}{0.5}$$

$$x = 24$$

solution is 24

(51.) $120 = \frac{x}{100}(80)$

$$100(120) = \overset{1}{\cancel{100}}\left(\frac{x}{\underset{1}{\cancel{100}}}\right)(80)$$

$$12000 = 80x$$

$$150 = x$$

solution is 150

(52.) $140 = \frac{x}{100}(2800)$

$$100(140) = \overset{1}{\cancel{100}}\left(\frac{x}{\underset{1}{\cancel{100}}}\right)(2800)$$

$$14000 = 2800x$$

$$5 = x$$

solution is 5

(53.) See problem (47.)

$$x = .05(34)$$

$$x = 1.7$$

solution is 1.7

(54.) See problem (48.)

$$x = .12(22)$$

$$x = 2.64$$

solution is 2.64

(55.) $x = .08(45)$

$$x = 3.6$$

tax is \$3.60

(56.) $x = .08(12)$

$$x = .96$$

tax is \$0.96

(57.) $.35x = 63$

$$\frac{.35x}{.35} = \frac{63}{.35}$$

$$x = 180$$

180 students

(58.) $.60x = 84$

$$\frac{.60x}{.60} = \frac{84}{.60}$$

$$x = 140$$

(59.) $12 = \frac{x}{100}(30)$

$$100(12) = \overset{1}{\cancel{100}}\left(\frac{x}{\underset{1}{\cancel{100}}}\right)(30)$$

$$1200 = 30x$$

$$40 = x$$

solution is 40

(60.) $42 = \frac{x}{100}(60)$

$$100(42) = \overset{1}{\cancel{100}}\left(\frac{x}{\underset{1}{\cancel{100}}}\right)(60)$$

$$4200 = 60x$$

$$70 = x$$

solution is 70

(61.) $180 = \frac{x}{100}(800)$

$$100(180) = \overset{1}{\cancel{100}}\left(\frac{x}{\underset{1}{\cancel{100}}}\right)(800)$$

$$18000 = 800x$$

$$22.5 = x$$

22.5%

(62.) $210 = \frac{x}{100}(750)$

$$100(210) = \overset{1}{\cancel{100}}\left(\frac{x}{\underset{1}{\cancel{100}}}\right)(750)$$

$$21000 = 750x$$

$$28 = x$$

28%

Additional Exercises

Solve the following equations and check each solution.

1. $x + 2 = 8$
2. $x + 3 = 7$
3. $x + 6 = -2$
4. $x - 7 = 12$
5. $5 = x + 1$
6. $3 = x - 7$
7. $x - 2 = -1$
8. $20 + x = 16$
9. $7 = 7 + x$
10. $15 + x = 14$
11. $12 = 12 + x$
12. $3x = 12$
13. $6x = 18$
14. $8 = 2x$
15. $9 = 9x$
16. $4x = -16$
17. $-10 = 5x$
18. $7x = 8$
19. $2x = -14$
20. $-3 = 2x$
21. $9x = 0$
22. $-8x = 16$
23. $-3x = 21$
24. $-7x = -7$
25. $-2x = -14$
26. $-x = 3$
27. $-x = -3$
28. $-x = 34$
29. $-x = -34$
30. $6 = -12x$
31. $\frac{1}{3}x = 4$
32. $5 = \frac{1}{5}x$
33. $\frac{1}{4}x = -3$
34. $\frac{1}{5}x = -5$
35. $-8 = \frac{1}{2}x$

36. $\frac{1}{20}x = 200$

37. $\frac{3}{4}x = 3$

38. $\frac{2}{3} = \frac{1}{3}x$

39. $16 = \frac{4}{5}x$

40. $-\frac{3}{7} = \frac{3}{7}x$

41. $\frac{4}{5}x = -4$

42. $\frac{21}{22}x = 0$

43. $4x = 4.8$

44. $-3x = -3.6$

45. $\frac{4x}{5} = 2.8$

46. $\frac{3x}{4} = 2.4$

47. $x = 0.04(28)$

48. $x = 0.14(24)$

49. $0.15x - 33$

50. $0.6x = 18$

51. $140 = \frac{x}{100}(20)$

52. $130 = \frac{x}{100}(2600)$

Do the following word problems.

53. What is 4% of 28?

54. What is 14% of 24?

55. The sales tax is 8%. Find the tax on a dress that costs \$34.

56. The sales tax is 8%. Find the tax on a shirt that costs \$7.50.

57. 40% of the freshman class at a college are men. If there are 320 men in the freshman class, how many students are there in that class?

58. 36% of the total number of students in a law school are women. If there are 180 women in the school, how many students are there in the school?

59. 14 is what percent of 42?

60. 55 is what percent of 80?

61. Jane Moore pays $140 a month for clothing, and her monthly income is $700. Clothing costs are what percent of her total income?

62. Peter Romeo pays $240 rent each month and earns $720 monthly. What percent of his earnings does he spend for rent?

Lesson 2

Advanced Linear Equations in One Variable

Some equations can get very complicated, but you can still use the basic idea: If two things are equal and you do the same thing to both of them, they are still equal. Remember that the easiest kinds of equations to solve are the ones that look like:

"x = number"

or "number = x"

EXAMPLE 1 Solve the equation $3x = 2x + 7$.

Let's head for "x = number." We want the x's on the left side of the equation and the numbers on the right side of the equation.

$$3x = 2x + 7$$

To get rid of the $2x$ on the right, add $-2x$ to both sides because $2x + (-2x) = 0$.

$$
\begin{array}{rrcl}
 & 3x & = & 2x + 7 \\
\text{Add} & \underline{-2x} & & \underline{-2x \quad\;\;} \\
 & x & = & 0 + 7
\end{array}
$$

which is the same as $x = 7$.

Since 7 makes $x = 7$ true, 7 also makes $3x = 2x + 7$ true. So the solution is 7. To check the solution, put 7 in for x in the original equation.

$$
\begin{array}{lll}
3x & = 2x + 7 & \\
3 \cdot 7 & = 2 \cdot 7 + 7 & \\
21 & = 14 + 7 & \\
21 & = 21 & \underline{\text{True!}}
\end{array}
$$

The solution 7 checks.

EXAMPLE 2 Solve the equation $4x - 3 = 3x + 1$.

Let's head for "x = number." We want the x's on the left side of the equation and the numbers on the right side of the equation.

$$
\begin{array}{lrcr}
 & 4x - 3 & = & 3x + 1 \\
\text{Add } -3x \text{ to both sides} & \underline{-\,3x \quad\;\;} & & \underline{-3x \quad\;\;} \\
 & x - 3 & = & 0 + 1 \\
 & x - 3 & = & 1 \\
\text{Add } +3 \text{ to both sides} & \underline{\quad +3} & & \underline{\quad +3} \\
 & x + 0 & = & 4 \\
 & x & = & 4
\end{array}
$$

So the solution is 4. Let's check it in the original equation.

$$
\begin{array}{llcrclll}
 & 4x & - & 3 & = & 3x & + 1 & \\
\text{Put 4 in for each } x & 4 \cdot 4 & - & 3 & = & 3 \cdot 4 & + 1 & \\
 & 16 & - & 3 & = & 12 & + 1 & \\
 & & & 13 & = & 13 & & \underline{\text{True!}}
\end{array}
$$

The solution 4 checks.

EXAMPLE 3 Solve the following equation: $5.3x - 4 = 2.3x + 2$.

Let's head for "x = number."

	$5.3x - 4$	$=$	$2.3x + 2$
Add $-2.3x$ to both sides	$-2.3x$		$-2.3x$
	$3x - 4$	$=$	$0 + 2$
Add $+4$ to both sides	$+ 4$		$+ 4$
	$3x$	$=$	6

Since the x is multiplied by 3, we divide both sides by 3.

$$\frac{3x}{3} = \frac{6}{3}$$

$$x = 2$$

So the solution is 2. Let's check it in the original equation.

$$5.3x - 4 = 2.3x + 2$$

Put 2 in for each x.

$$5.3(2) - 4 = 2.3(2) + 2$$
$$10.6 - 4 = 4.6 + 2$$
$$6.6 = 6.6 \qquad \underline{\text{True!}}$$

The solution 2 checks.

EXAMPLE 4 Solve the following equation: $x + 3x - 6.2 = 2x + 7.8$.

In this equation the left side can be simplified.

$$4x - 6.2 = 2x + 7.8$$

Let's head for "x = number."

	$4x - 6.2$	$=$	$2x + 7.8$
Add $-2x$ to both sides	$-2x$		$-2x$
	$2x - 6.2$	$=$	7.8
Add $+6.2$ to both sides	$+ 6.2$	$=$	$+ 6.2$
	$2x$	$=$	14

Since x is multiplied by 2, we divide both sides by 2.

$$\frac{2x}{2} = \frac{14}{2}$$

$$x = 7$$

So the solution is 7. Check in the original equation.

EXAMPLE 5 Solve the equation $x - 1 = x$.

This one is tricky; let's head for "x = number."

$$\begin{array}{lrcr} & x - 1 & = & x \\ \text{Add } -x \text{ to both sides} & -x & & -x \\ \hline & -1 & = & 0 \end{array}$$

But the equation $-1 = 0$ is false. So, there is nothing that can make $x - 1 = x$ true, and we say the equation has no solution.

EXAMPLE 6 Solve the equation $3x + 7 = 2x + 7$.

Let's head for "x = number."

$$\begin{array}{lrcr} & 3x + 7 & = & 2x + 7 \\ \text{Add } -2x \text{ to both sides} & -2x & & -2x \\ \hline & x + 7 & & 7 \\ \text{Add } -7 \text{ to both sides} & -7 & & -7 \\ \hline & x & = & 0 \end{array}$$

So the solution is 0.

Any letter or symbol can be used as a variable. In the next few examples we use letters other than x for the variable. Notice that the solution methods do not change.

EXAMPLE 7 Solve the equation $3(2y - 1) = 1 + 8y$.

$$3(2y - 1) = 1 + 8y$$

is the same as $6y - 3 = 1 + 8y$

Let's head for "number = y."

$$\begin{array}{lrcr} & 6y - 3 & = & 1 + 8y \\ \text{Add } -6y \text{ to both sides} & -6y & & -6y \\ \hline & -3 & = & 1 + 2y \\ \text{Add } -1 \text{ to both sides} & -1 & & -1 \\ \hline & -4 & = & 2y \end{array}$$

We want "number = y" so divide both sides by 2.

$$\frac{-4}{2} = \frac{2y}{2}$$

$$-2 = y$$

So the solution is -2. Let's check it in the original equation.

$$3(2y - 1) = 1.8y$$

Put in -2 for each y.

$$\begin{aligned} 3(2(-2) - 1) &= 1 + 8(-2) \\ 3(-4 - 1) &= 1 - 16 \\ 3(-5) &= -15 \\ -15 &= -15 \qquad \underline{\text{True!}} \end{aligned}$$

It checks.

EXAMPLE 8 Solve the equation $2 - (z + 3) = -2(z + 4)$.

First simplify.

$$\begin{aligned} 2 - z - 3 &= -2z - 8 \\ -z - 1 &= -2z - 8 \end{aligned}$$

Let's head for "number = z."

$$\begin{array}{lrcl} & -z - 1 &=& -2z - 8 \\ \text{Add } +z \text{ to both sides} & +z & & +\ z \\ \hline & -1 &=& -\ z - 8 \\ \text{Add } +8 \text{ to both sides} & +8 & & +\ 8 \\ \hline & +7 &=& -\ z \end{array}$$

But we want "number = z" so we multiply both sides by -1.

$$+7(-1) = -z(-1)$$

$$-7 = z$$

So the solution is -7.

EXAMPLE 9 Solve the following equation: $\dfrac{x}{2} = 6 - \dfrac{x}{4}$

To get rid of the fractions, multiply both sides by 4 (the lowest common denominator).

$$4\left(\frac{x}{2}\right) = 4\left(6 - \frac{x}{4}\right)$$

$$4\left(\frac{x}{2}\right) = 4(6) - 4\left(\frac{x}{4}\right)$$

$$2x = 24 - x$$

Let's head for "x = number."

$$\begin{array}{lrcl} & 2x &=& 24 - x \\ \text{Add } x \text{ to both sides} & +\ x & & +\ x \\ \hline & 3x &=& 24 \\ & x &=& 8 \end{array}$$

So the solution is 8.

EXAMPLE 10 Solve the following equation: $\frac{2r}{3} - \frac{r}{2} = 10 - \frac{r}{4}$

To get rid of the fractions, multiply both sides by 12 (the lowest common denominator).

$$12\left(\frac{2r}{3} - \frac{r}{2}\right) = 12\left(10 - \frac{r}{4}\right)$$

$$12\left(\frac{2r}{3}\right) - 12\left(\frac{r}{2}\right) = 12(10) - 12\left(\frac{r}{4}\right)$$

$$8r - 6r = 120 - 3r$$

Let's simplify.

$$2r = 120 - 3r$$

Let's head for "r = number."

$$\begin{array}{lrcl} & 2r & = & 120 - 3r \\ \text{Add } +3r \text{ to both sides} & +3r & & +\ 3r \\ \hline & 5r & = & 120 \\ \text{Divide both sides by 5} & \frac{5r}{5} & = & \frac{120}{5} \\ & r & = & 24 \end{array}$$

So the solution is 24. Let's check it in the original equation.

$$\frac{2r}{3} - \frac{r}{2} = 10 - \frac{r}{4}$$

Put in 24 for <u>each</u> r.

$$\frac{2(24)}{3} - \frac{24}{2} = 10 - \frac{24}{4}$$

$$\frac{48}{3} - 12 = 10 - 6$$

$$16 - 12 = 4$$

$$4 = 4 \qquad \underline{\text{True!}}$$

The solution 24 checks.

EXAMPLE 11 A 15.6-meter pole is cut into two pieces so that one piece is twice as long as the other. How long are each of the two pieces?

We let p stand for the length of the shorter piece. The longer piece is two times the length of the shorter, so the length of the longer piece is $2p$.

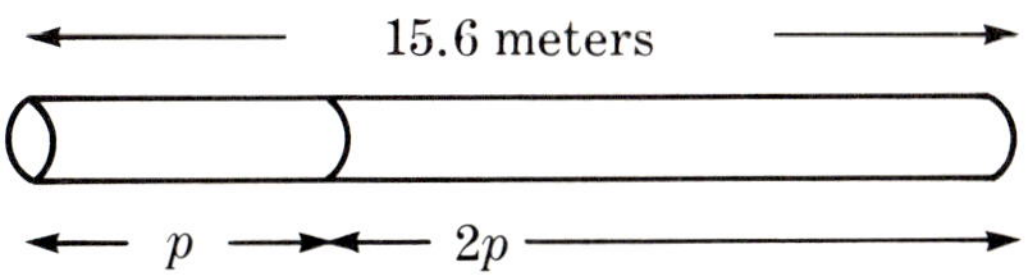

length of pole = short piece + long piece

$$\begin{aligned} 15.6 &= p + 2p \\ 15.6 &= 3p \\ 5.2 &= p \end{aligned}$$

So the length of the shorter piece, which is p, is 5.2 meters. The length of the longer piece, which is $2p$, is 2(5.2) or 10.4 meters.

Note: $5.2 + 10.4 = 15.6$.

EXAMPLE 12 A pole is standing upright in a pond. 20% of the pole is in the ground, 40% is in the water and 10 feet of the pole are in the air. How long is the pole?

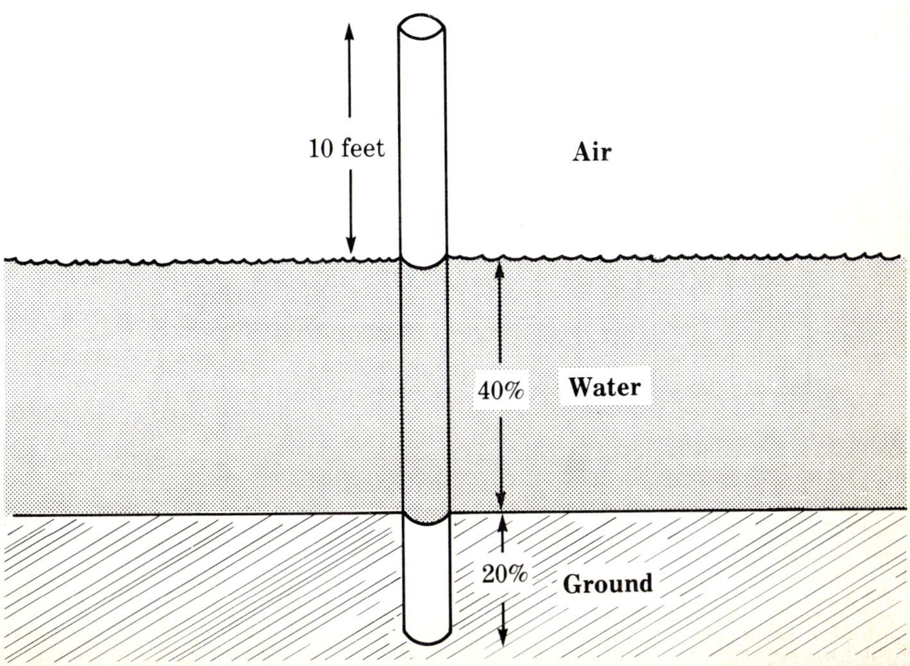

We let p stand for the length of the pole.

$$\begin{array}{ccccccc} \text{in ground} & + & \text{in water} & + & \text{in air} & = & \text{length of pole} \\ \downarrow & & \downarrow & & \downarrow & & \downarrow \\ 0.20p & + & 0.40p & + & 10 & = & p \end{array}$$

$$0.60p + 10 = p$$

Add $-0.60p$ to both sides

$$\begin{array}{rcl} 0.60p + 10 & = & p \\ -0.60p & & -0.60p \\ \hline 10 & = & p - 0.60p \end{array}$$

Note: $p - 0.60p$ can be written as $1.00p - 0.60p$ or $0.40p$.

$$10 = 0.40p$$

$$\frac{10}{0.40} = \frac{\overset{1}{\cancel{0.40}}p}{\underset{1}{\cancel{0.40}}}$$

$$25 = p$$

So the length of the pole, which is p, is 25 feet. To check our answer, we calculate the three lengths and add them.

20% of the pole is in the ground, and 0.20(25) is 5 ft

40% of the pole is in the water, and 0.40(25) is 10 ft

10 feet of the pole is in the air

5 ft + 10 ft + 10 ft = 25 ft

Exercises

Solve and check.

1. $5x = 4x + 3$
2. $7x = 6x - 2$
3. $2x + 5 = x$
4. $5y - 3 = 7$
5. $3x - 6 = 5x$
6. $7x + 3 = -18$
7. $6x + 2 = 5x + 4$
8. $2x + 3.1 = 3x - 7.4$
9. $5a - 5 = 2a + 4$
10. $5x + 3 = 13 - 5x$
11. $8 - 7b = 2b - 1$
12. $-5 - 3x = -3 - 2x$
13. $x + 2x = 6.6$
14. $5x - 2x = 9 - 3$
15. $6c - 8 = 2c - 4 + 2.4$
16. $x + 2x + 1 = 2x + 4$
17. $10 + 7z + 5 = 5z + 17$
18. $3 - 5x + 1 = -4.6x$
19. $5.2x - 3 = 1.2x + 7$
20. $x + 3 = x$
21. $w + 3 = 3$
22. $2r + 8 = 5r + 8$
23. $3(x + 2) = 9$
24. $3x - 2 = 2(5 + x)$
25. $6(3y - 1) = -42$
26. $2(5x - 1) = 7(x - 2)$
27. $2(x - 3) = -(x - 12)$
28. $-3(y - 1) = -2(y - 4)$
29. $-5(2x + 5) = 2 + (x + 6)$
30. $\frac{x}{2} = \frac{11}{2}$
31. $\frac{a}{2} + \frac{a}{4} = 6$
32. $\frac{x}{3} - \frac{x}{5} = 2$
33. $-\frac{x}{2} + 5 = \frac{x}{3}$
34. $\frac{x}{4} + 1 = \frac{x}{2} - \frac{x}{6}$

Do the following word problems.

35. In a book store, the day's receipts of \$253.80 included 8% sales tax. How much sales tax was collected?

36. Anthony invests some money at 6% and three times as much money at 8%. If the yearly income from the two investments is \$300, how much is invested at each rate?

37. A factory employs 5 engineers and 60 mechanics. The weekly wage of an engineer is \$100 more than the weekly wage of a mechanic. If the total weekly payroll is \$20,000, how much does an engineer earn each week and how much does a mechanic earn each week?

38. 280 feet of fencing is needed to completely enclose a rectangular dog run. The width is four tenths of the length. Find the width and length of the dog run.

Answers to Exercises

1. $$\begin{aligned} 5x &= 4x + 3 \\ -4x & \quad -4x \\ x &= 0 + 3 \\ x &= 3 \end{aligned}$$

2. $$\begin{aligned} 7x &= 6x - 2 \\ -6x & \quad -6x \\ x &= 0 - 2 \\ x &= -2 \end{aligned}$$

3. $$\begin{aligned} 2x + 5 &= x \\ -x \quad & \quad -x \\ x + 5 &= 0 \\ -5 & \quad -5 \\ x + 0 &= -5 \\ x &= -5 \end{aligned}$$

4. $$\begin{aligned} 5y - 3 &= 7 \\ +3 & \quad +3 \\ 5y + 0 &= 10 \\ \frac{5y}{5} &= \frac{10}{5} \\ y &= 2 \end{aligned}$$

5. $$\begin{aligned} 3x - 6 &= 5x \\ -3x \quad & \quad -3x \\ 0 - 6 &= 2x \\ \frac{-6}{2} &= \frac{2x}{2} \\ -3 &= x \\ \text{or } x &= -3 \end{aligned}$$

6. $$\begin{aligned} 7x + 3 &= -18 \\ -3 & \quad -3 \\ 7x + 0 &= -21 \\ \frac{7x}{7} &= \frac{-21}{7} \\ x &= -3 \end{aligned}$$

7. $$\begin{aligned} 6x + 2 &= 5x + 4 \\ -5x \quad & \quad -5x \\ x + 2 &= 0 + 4 \\ -2 & \quad -2 \\ x + 0 &= 2 \\ x &= 2 \end{aligned}$$

8. $$\begin{aligned} 2x + 3.1 &= 3x - 7.4 \\ -2x \quad & \quad -2x \\ 0 + 3.1 &= x - 7.4 \\ +7.4 & \quad +7.4 \\ 10.5 &= x \\ \text{or } x &= 10.5 \end{aligned}$$

9. $$\begin{aligned} 5a - 5 &= 2a + 4 \\ -2a \quad & \quad -2a \\ 3a - 5 &= 0 + 4 \\ +5 & \quad +5 \\ 3a + 0 &= 9 \\ \frac{3a}{3} &= \frac{9}{3} \\ a &= 3 \end{aligned}$$

(10.)
$$5x + 3 = 13 - 5x$$
$$\underline{+5x} \qquad \underline{+5x}$$
$$10x + 3 = 13 + 0$$
$$\underline{-3} \qquad \underline{-3}$$
$$10x + 0 = 10$$
$$\frac{10x}{10} = \frac{10}{10}$$
$$x = 1$$

(11.)
$$8 - 7b = 2b - 1$$
$$\underline{+7b} \qquad \underline{+7b}$$
$$8 + 0 = 9b - 1$$
$$\underline{+1} \qquad \underline{+1}$$
$$9 + 0 = 9b + 0$$
$$\frac{9}{9} = \frac{9b}{9}$$
$$1 = b$$
or $b = 1$

(12.)
$$-5 - 3x = -3 - 2x$$
$$\underline{+3x} \qquad \underline{+3x}$$
$$-5 + 0 = -3 + x$$
$$\underline{+3} \qquad \underline{+3}$$
$$-2 + 0 = 0 + x$$
$$-2 = x$$
or $x = -2$

(13.)
$$x + 2x = 6.6$$
$$3x = 6.6$$
$$x = 2.2$$

(14.)
$$5x - 2x = 9 - 3$$
$$3x = 6$$
$$x = 2$$

(15.)
$$6c - 8 = 2c - 4 + 2.4$$
$$6c - 8 = 2c - 1.6$$
$$\underline{+8} \qquad \underline{+8}$$
$$6c = 2c + 6.4$$
$$\underline{-2c} \qquad \underline{-2c}$$
$$4c = 6.4$$
$$c = 1.6$$

(16.)
$$x + 2x + 1 = 2x + 4$$
$$\underline{-1} \qquad \underline{-1}$$
$$3x = 2x + 3$$
$$\underline{-2x} \qquad \underline{-2x}$$
$$x = 3$$

(17.)
$$10 + 7z + 5 = 5z + 17$$
$$\underline{-5z} \qquad \underline{-5z}$$
$$15 + 2z = 17$$
$$\underline{-15} \qquad \underline{-15}$$
$$2z = +2$$
$$z = 1$$

18. $3 - 5x + 1 = -4.6x$

$+5x \qquad +5x$

$4 = 0.4x$

$\frac{4}{0.4} = \frac{0.4x}{0.4}$

$10 = x$

19. $5.2x - 3 = 1.2x + 7$

$+3 \qquad +3$

$5.2x = 1.2x + 10$

$-1.2x \qquad -1.2x$

$4x = +10$

$x = \frac{10}{4}$

$x = 2.5$

20. $x + 3 = x$

$-x \qquad -x$

$3 = 0$ false

no solution!

21. $w + 3 = 3$

$-3 \qquad -3$

$w = 0$

22. $2r + 8 = 5r + 8$

$-8 \qquad -8$

$2r = 5r$

$-2r \qquad -2r$

$0 = 3r$

$0 = r$

23. $3(x + 2) = 9$

$3x + 6 = 9$

$-6 \qquad -6$

$3x = 3$

$x = 1$

24. $3x - 2 = 2(5 + x)$

$3x - 2 = 10 + 2x$

$+2 \qquad +2$

$3x = 12 + 2x$

$-2x \qquad -2x$

$x = 12$

25. $6(3y - 1) = -42$

$18y - 6 = -42$

$+6 \qquad +6$

$18y = -36$

$y = -2$

26. $2(5x - 1) = 7(x - 2)$

$10x - 2 = 7x - 14$

$+2 \qquad +2$

$10x = 7x - 12$

$-7x \qquad -7x$

$3x = -12$

$x = -4$

(27.) $2(x-3) = -(x-12)$
$2x-6 = -x+12$
$+6 \quad +6$
$2x = -x+18$
$+x \quad +x$
$3x = 18$
$x = 6$

(28.) $-3(y-1) = -2(y-4)$
$-3y+3 = -2y+8$
$+3y \quad +3y$
$+3 = y+8$
$-8 \quad -8$
$-5 = y$
or $y = -5$

(29.) $-5(2x+5) = 2+(x+6)$
$-10x-25 = 2+x+6$
$+10x \quad +10x$
$-25 = 8+11x$
$-8 \quad -8$
$-33 = 11x$
$-3 = x$
or $x = -3$

(30.) $\frac{x}{2} = \frac{11}{2}$
$2\left(\frac{x}{2}\right) = \left(\frac{11}{2}\right)2$
$x = 11$

(31.) $\frac{a}{2}+\frac{a}{4} = 6$ LCD is 4
$4\left(\frac{a}{2}\right)+4\left(\frac{a}{4}\right) = 6(4)$
$2a+a = 24$
$3a = 24$
$a = 8$

(32.) $\frac{x}{3}-\frac{x}{5} = 2$ LCD is 15
$15\left(\frac{x}{3}\right)-15\left(\frac{x}{5}\right) = 15(2)$
$5x-3x = 30$
$2x = 30$
$x = 15$

(33.) $-\frac{x}{2}+5 = \frac{x}{3}$ LCD is 6
$6\left(-\frac{x}{2}\right)+6(5) = 6\left(\frac{x}{3}\right)$
$-3x+30 = 2x$
$+3x \quad +3x$
$30 = 5x$
$6 = x$
or $x = 6$

(34.) $\frac{x}{4} + 1 = \frac{x}{2} - \frac{x}{6}$ LCD is 12

$$12\left(\frac{x}{4}\right) + 12(1) = 12\left(\frac{x}{2}\right) - 12\left(\frac{x}{6}\right)$$

$$3x + 12 = 6x - 2x$$
$$3x + 12 = 4x$$
$$\underline{-3x \qquad\qquad -3x}$$
$$12 = x$$

or $x = 12$

(35.) Let s stand for amount of sales.

Then $0.08s$ would be the amount of the sales tax.

Sales + Sales tax = Receipts

$$s + 0.08s = 253.80$$
$$1.08s = 253.80$$
$$\frac{1.08s}{1.08} = \frac{253.80}{1.08}$$
$$s = 235$$

The amount of sales, which is s, is \$235. The sales tax is $0.08s(235)$ or \$18.80.

Sales tax collected is \$18.80

(36.) Let x be amount invested at 6%

Then $3x$ is amount invested at 8%

$$0.06x + 0.08(3x) = 300$$
$$0.06x + 0.24x = 300$$
$$0.30x = 300$$

$$\frac{0.30x}{0.30} = \frac{300}{0.30}$$

$$x = 1000$$

\$1000 invested at 6%

\$3000 invested at 8%

(37.) Let m stand for each mechanic's weekly wage. Then engineer's wage is $m + 100$.

payroll = Total Engineers' wages + total mechanics' wages.

$$20000 = 5(m+100) + 60m$$
$$20000 = 5m + 500 + 60m$$
$$20000 = 65m + 500$$
$$19500 = 65m$$

$$\frac{19500}{65} = \frac{65m}{65}$$
$$300 = m$$

Mechanic earns \$300 a week
Engineer earns \$400 a week

(38.) Let l stand for length

$0.4l$ stands for width

length is l

width is $0.4l$  width is $0.4l$

length is l

perimeter is distance around the rectangle

$$280 = l + 0.4l + l + 0.4l$$

$$280 = 2.8l$$

$$\frac{280}{2.8} = \frac{2.8l}{2.8}$$

$$100 = l$$

length is 100 feet

width is $0.4(100)$ or 40 feet.

Additional Exercises

Solve and check.

1. $6x = 5x + 2$
2. $9x = 8x - 1$
3. $4x - 6 = 3x$
4. $4y - 5 = 7$
5. $7x - 5 = 2x$
6. $6x + 4 = -14$
7. $2x + 1 = x - 2$
8. $5x + 4.1 = 4x - 7.3$
9. $3b - 6 = b + 2$
10. $15 + 7x = 1 - 7x$
11. $7 - 2x = 3x - 8$
12. $-4 - 2x = -2 - x$
13. $3x + x = 12$
14. $4x - x = 9 - 3$
15. $5x - 7 = 2x - 10.6$
16. $x + 5x - 2 = 3x + 7$
17. $a + 6 - 3 = 3a - 9$
18. $8 - 4x - 3 = -2x + 5$
19. $6.7x - 1 = 4.9x + 8$
20. $x + 4 = x$
21. $y + 2 = 2$
22. $3z + 5 = 7z + 5$
23. $5(x + 1) = 10$
24. $4x - 2 = 3(6 + x)$
25. $5(r - 3) = -10$
26. $3(x - 2) = 2(x + 1)$
27. $4(x - 1) = -(x - 11)$
28. $2(3y - 2) = 3(4y - 2) - 16$
29. $5(x + 3) = -2(x - 11)$
30. $3 - (x + 4) = -3(x + 5)$
31. $\frac{a}{2} + \frac{a}{3} = 5$
32. $\frac{x}{3} + \frac{x}{6} = 3$
33. $\frac{2x}{3} + \frac{x}{4} = \frac{11}{12}$
34. $\frac{w}{5} + \frac{1}{10} = \frac{w}{4}$
35. $\frac{x}{6} - \frac{3x}{4} = 1 - \frac{2}{3}x$

Do the following word problems.

36. In a shoe store the day's receipts of $314.70 included the 8% sales tax. How much sales tax was collected?

37. Susan invests some money at 5% and four times as much money at 6%. If the yearly income from the two investments is $290, how much is invested at each rate?

38. A rectangular playground is completely enclosed with 480 feet of fencing. If the width is six tenths of the length, find the width and length of the playground.

Lesson 3

Linear Equations in Two Variables

The sentence $x + y = 5$ is an equation. It is not true and it is not false. If 3 is put in place of x, the equation $3 + y = 5$ is still neither true nor false. However, if 3 is put in place of x, and 2 is put in place of y, then the sentence becomes true; $3 + 2 = 5$. If 4 is put in place of x and 2 is put in place of y, then the sentence becomes false; $4 + 2 = 5$.

In $x + y = 5$ we must put in two numbers (one for x and one for y) to see if the sentence becomes true or false. Finding the pairs of numbers that make this kind of equation true is called **solving** the equation.

Let's write some of the solutions of the equation $x + y = 5$ in a table:

x	y
+3	+2
+2	+3
+1	+4
0	+5
−1	+6
−2	+7

We can write the first solution as (+3, +2). We always write the numbers in (x, y) order with parentheses; that is, the "x" is the first number and the "y" is the second. The first number is called the **x-coordinate** of the solution, and the second number is called the **y-coordinate** of the solution.

Another way of showing solutions to this kind of an equation is called **graphing**. When we graph solutions, we draw two number lines crossing each other as in Figure 1.

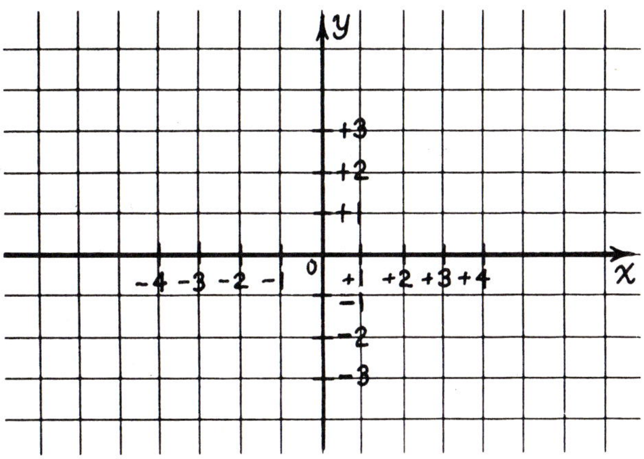

Figure 1

The number line pointing to the right is called the **x-axis**, and the number line pointing up is called the **y-axis**. The purpose of the arrowheads is to remind us of the positive direction of each of the number lines. The point where the x-axis and the y-axis cross is called the **origin**. See Figure 2.

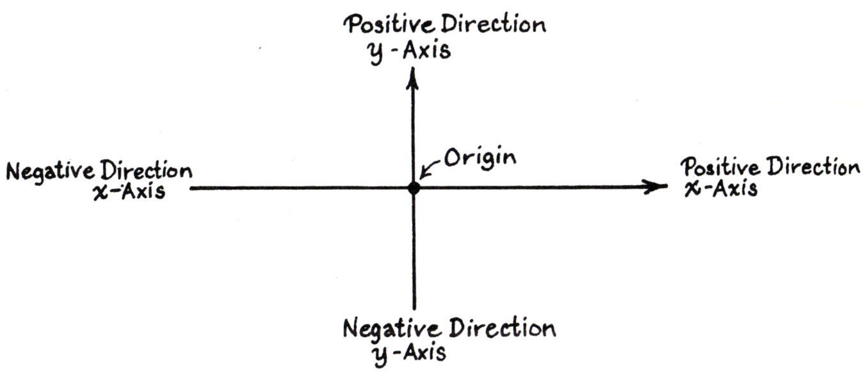

Figure 2

We will graph (plot) a solution of $x + y = 5$ as a single point. The solution we will use is $(+3, +2)$. Start at the origin where x is zero and y is zero. The origin is $(0, 0)$. Since $+3$, the x-coordinate, is a positive number, move three units in the positive x direction (to the right) along the x-axis. Since $+2$, the y-coordinate, is positive, continue in the positive y direction (up) two units as shown in Figure 3. Make a dot. This dot is the graph of $(+3, +2)$.

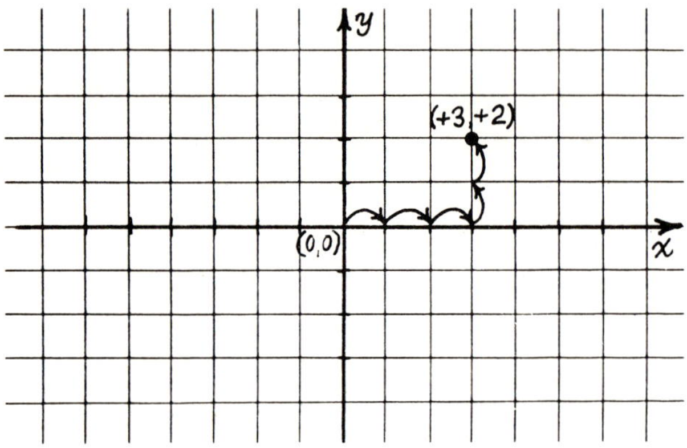

Figure 3

Let's graph a few of the other solutions of $x + y = 5$. The next solution in the table is $(+2, +3)$. Watch it! This is not the same solution as the one we just plotted. This time x is 2 and y is 3. Start at the origin. Since $+2$, the x-coordinate, is a positive number, move two units in the positive x direction (to the right) along the x-axis. Since $+3$, the y-coordinate, is positive, continue in the positive y direction (up) three units as shown in Figure 4. Make a dot. This dot is the graph of $(+2, +3)$.

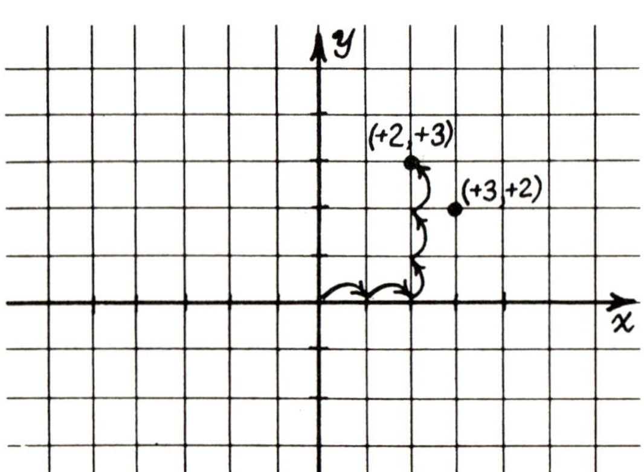

Figure 4

The point (+1, +4) is another solution. To graph this point, start at the origin and move one unit in the positive direction of the x-axis (to the right), and then four units in the positive y-direction (up). Make a dot. This dot is the graph of (+1, +4). See Figure 5.

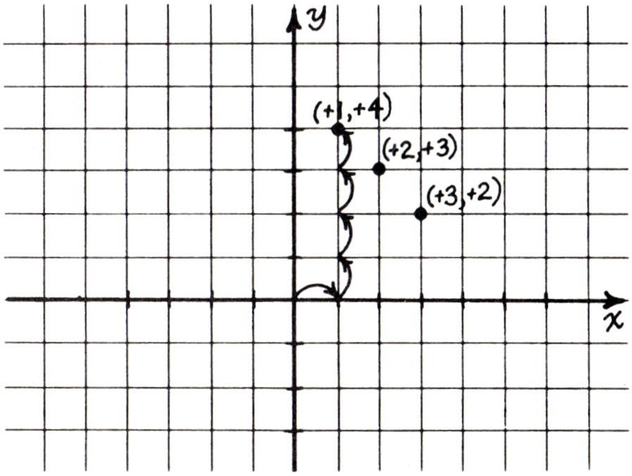

Figure 5

To graph the solution (0, +5), start at the origin and move 0 units along the x-axis (that is, stay at the origin), and then move 5 units up, as in Figure 6.

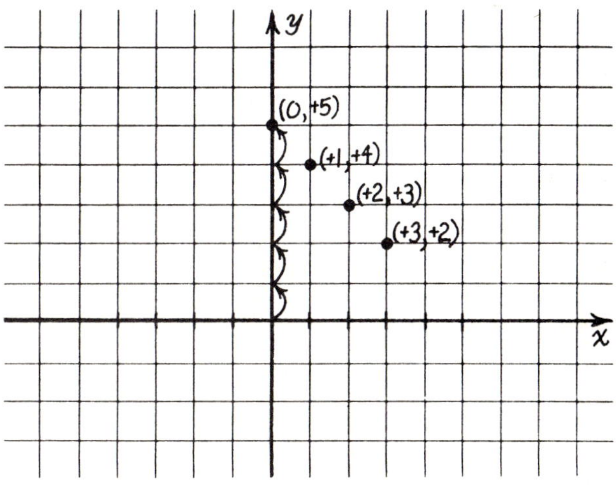

Figure 6

To graph $(-1, +6)$, start at the origin. Since -1, the x-coordinate, is negative, move one unit to the left (in the negative direction of the x-axis) and then six units in the positive y-direction (up). Make a dot. This dot is the graph of $(-1, +6)$. See Figure 7.

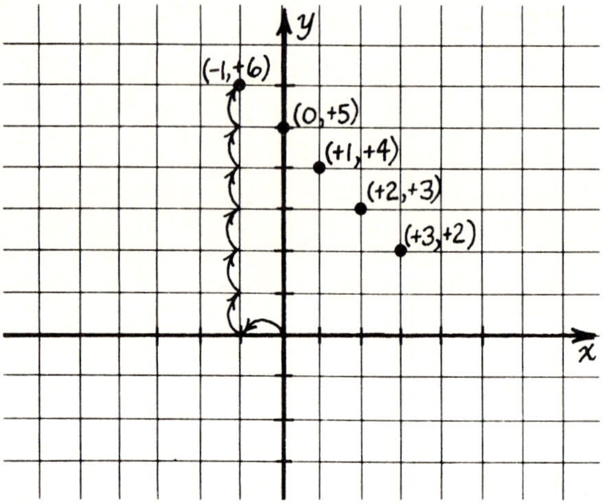

Figure 7

If all of the solutions in the table were graphed, Figure 8 would be the result.

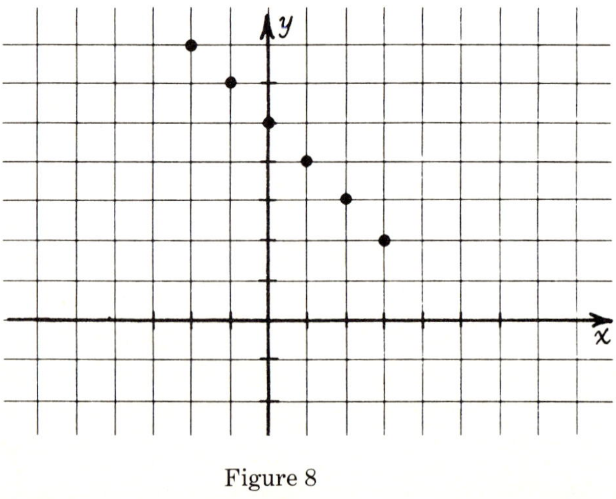

Figure 8

Notice how nicely these points line up! If you think up more solutions to the equation, they would line up with those already plotted. In fact, all the solutions of the equation will line up with those solutions already plotted.

Another way of saying this is: All the solutions of $x + y = 5$ lie on the same straight line. Plotting solutions like $\left(+\frac{1}{2}, +4\frac{1}{2}\right)$ and (+1.5, +3.5) will help "fill in" the line. See Figure 9.

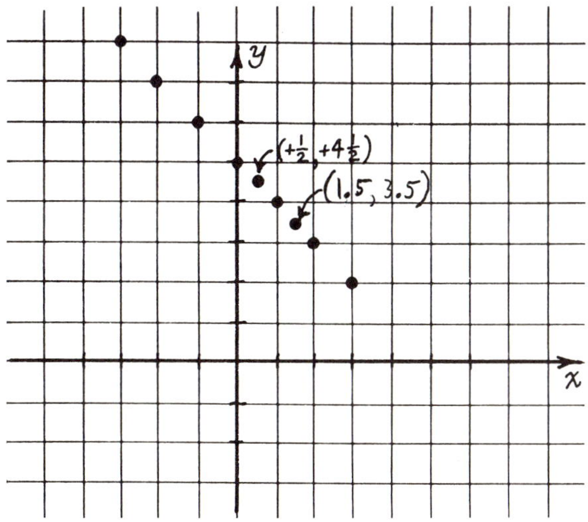

Figure 9

Thus, if all the solutions of the equation $x + y = 5$ could be plotted, the line would be infinitely long. Since we don't have enough paper, Figure 10 shows a small part of the line. An equation whose graph is a straight line is called a **linear equation**. Notice the word "line" in the word "linear."

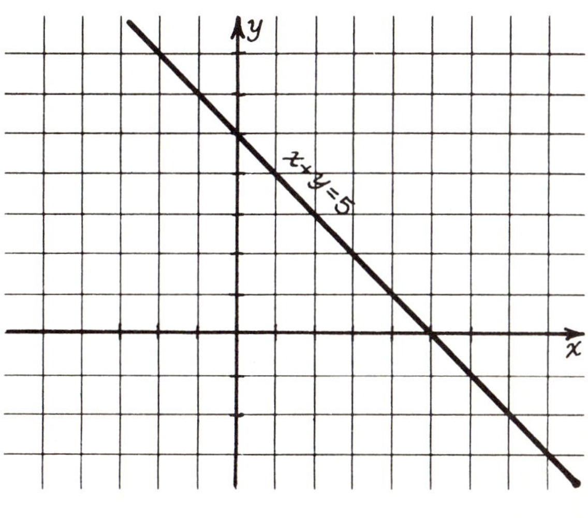

Figure 10

Any point on this line is a solution to the equation. For example, the point $(+7, -2)$ is on the line. See Figure 11.

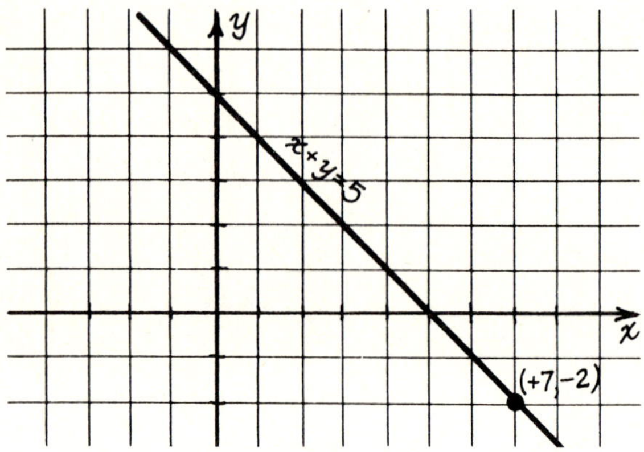

Figure 11

To show that $(+7, -2)$ is a solution of the equation $x + y = 5$, plug in $+7$ for x and plug in -2 for y.

$$\begin{aligned} x + y &= 5 \\ (+7) + (-2) &= 5 \\ 7 - 2 &= 5 \\ 5 &= 5 \quad \text{True!} \end{aligned}$$

So $(+7, -2)$ is a solution of the equation $x + y = 5$.

Any point not on the line is not a solution of the equation. For example, the point $(+4, +3)$ is not on the line. See Figure 12.

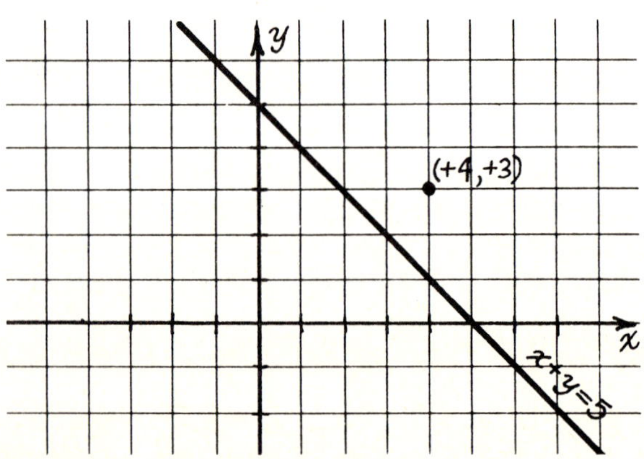

Figure 12

To show that (+4, +3) is not a solution, plug +4 in for x, and plug +3 in for y in the equation $x + y = 5$.

$$\begin{aligned} x + y &= 5 \\ +4 + 3 &= 5 \\ 7 &= 5 \qquad \underline{\text{False!}} \end{aligned}$$

So (+4, +3) is not a solution to the equation $x + y = 5$.

When graphing the solutions of a linear equation, we need to plot only two points, since we can draw the line using those two points. However, we can check that we haven't made any mistakes by plotting a third point which must also be on the line.

EXAMPLE 1 Graph the solutions of $x + y = 4$.

First determine any three solutions of the equation and write them in a table.

x	y
0	+4
+2	+2
+4	0

Plot the points as in Figure 13.

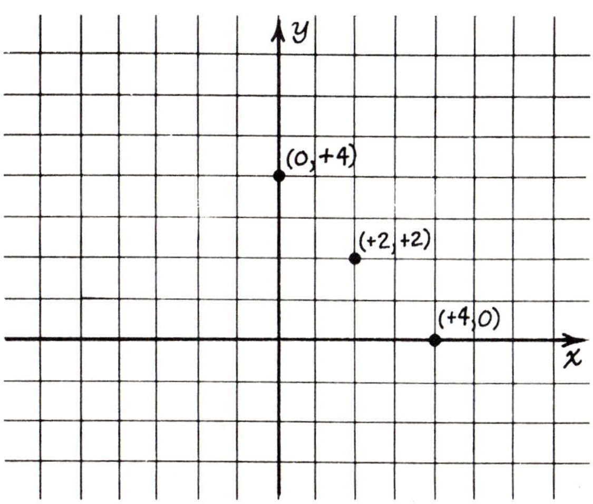

Figure 13

The straight line through these points is the graph of $x + y = 4$. See Figure 14.

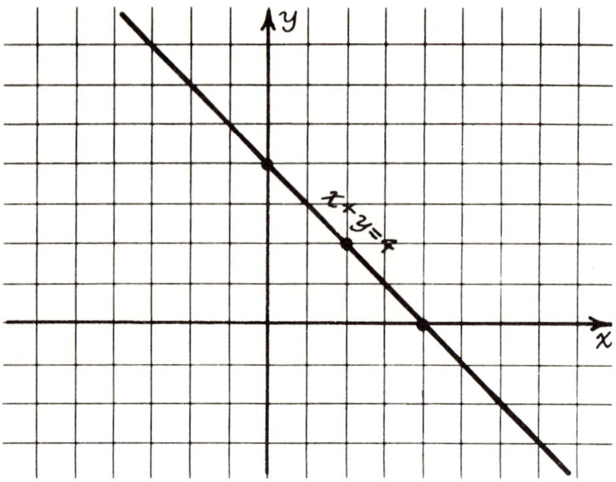

Figure 14

EXAMPLE 2 Graph the solutions of $x - y = 0$.

Three solutions are shown in the table below.

x	y
4	4
2	2
0	0

These points are plotted in Figure 15. (Notice that the point (0, 0) is the origin.)

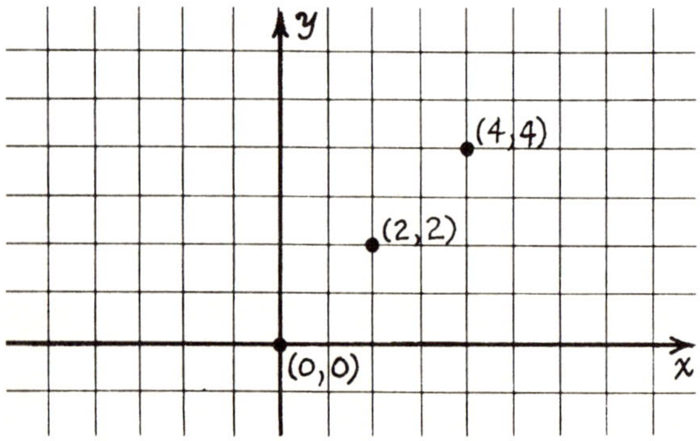

Figure 15

The straight line through these points is the graph of $x - y = 0$. See Figure 16.

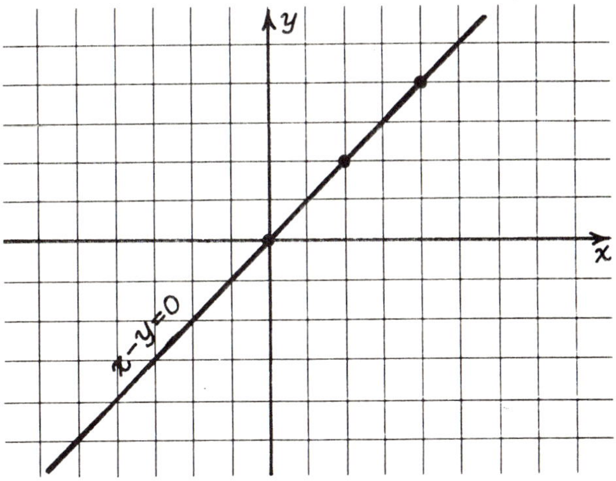

Figure 16

EXAMPLE 3 Graph the solutions of $y = 3$.

Any point whose y-coordinate is 3 is a solution. Since there is no x in the equation, the x-coordinate of a solution can be any number. Three solutions are shown in the table below.

x	y
0	3
2	3
−4	3

These points are plotted in Figure 17.

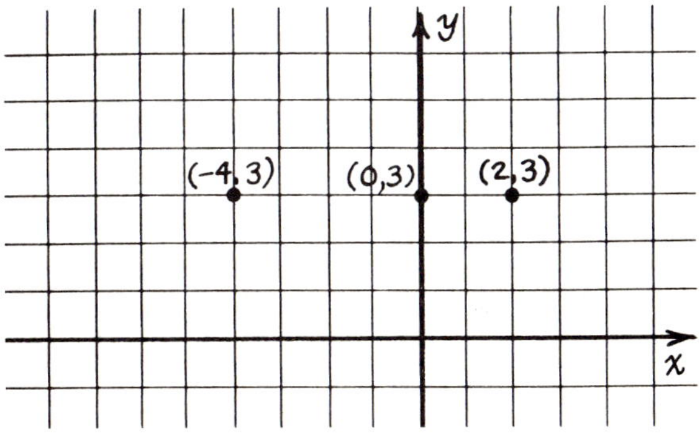

Figure 17

The straight line through these points is the graph of $y = 3$. See Figure 18.

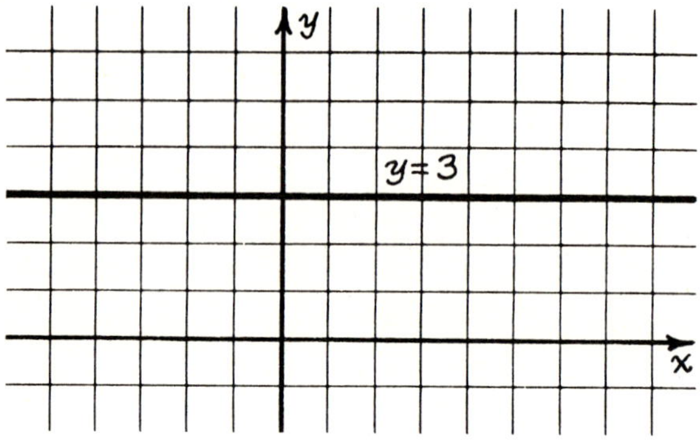

Figure 18

In example 3 we graphed the solutions of the equation $y = 3$ and got a horizontal line. Every point on that line had 3 as its y-coordinate. In general, the graph of every equation of the form $y = k$ (where k is some number) is a horizontal line and every point on that line will have k as its y-coordinate.

EXAMPLE 4 Graph the solutions of $x = -4$.

Any point whose x-coordinate is -4 is a solution. Since there is no y in the equation, the y-coordinate of a solution can be any number. Three solutions are shown in the table which follows.

x	y
-4	0
-4	3
-4	-2

These points are plotted in Figure 19.

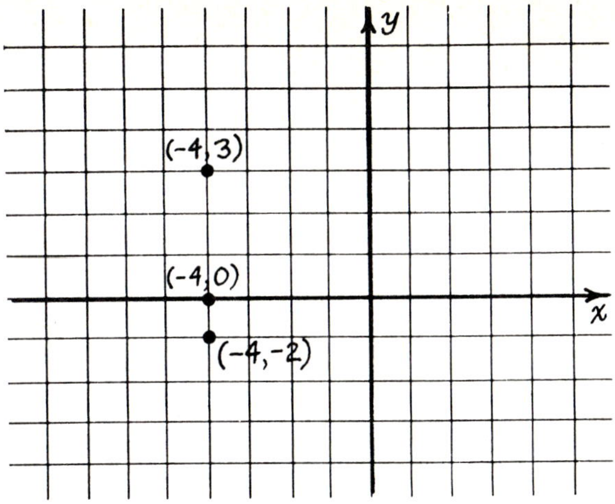

Figure 19

The straight line through these points is the graph of $x = -4$. See Figure 20.

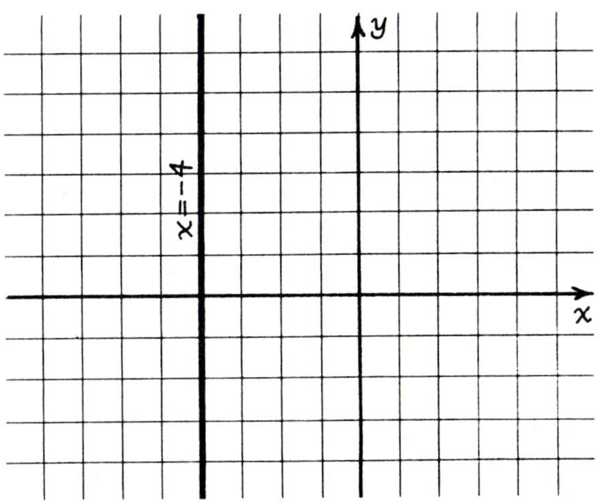

Figure 20

In example 4 we graphed the solutions of the equation $x = -4$ and got a vertical line. Every point on that line had -4 as its x-coordinate. In general, the graph of every equation of the form $x = k$ (where k is some number) is a vertical line and every point on that line will have k as its x-coordinate.

Exercises

1. Graph the following points and identify each point with its x-coordinate and y-coordinate.

(2, 3)	(0, 0)	$(-1\frac{1}{2}, 0)$
(3, 2)	(−2, −5)	(0, 2.5)
(−1, 4)	(−4, 0)	
(4, −1)	(0, −2)	
(3, 0)	(1, −5)	

2. Determine the coordinates of the following points.

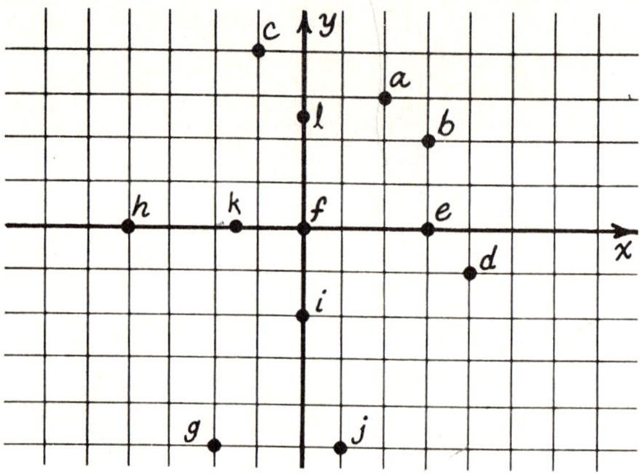

Graph the solutions of each of the following equations, using a separate set of axes for each. Write the equation along each line graph.

3. $x + y = 8$
4. $x + y = 3$
5. $x - y = 5$
6. $y = 2x$
7. $y = -3x$
8. $x + y = 1$
9. $y - x = 0$
10. $y = x$
11. $y - x = 5$
12. $x + y = -1$
13. $x - y = -2$
14. $y - x = -3$
15. $y = -2$
16. $2y = 6$
17. $x = 4$
18. $x = 0$

Answers to Exercises

1.

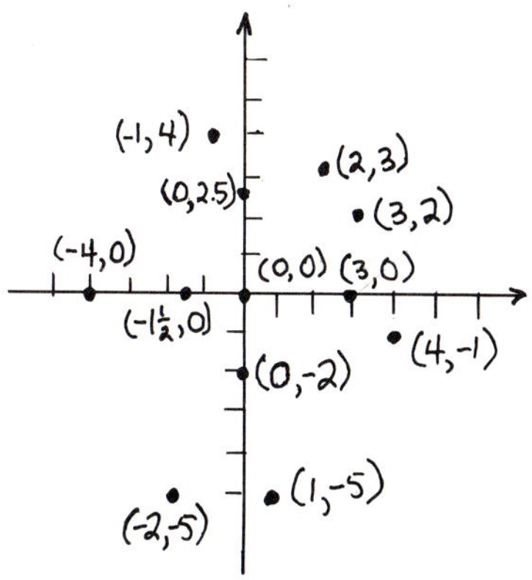

2.

a - (2,3)	g - (-2,-5)
b - (3,2)	h - (-4,0)
c - (-1,4)	i - (0,-2)
d - (4,-1)	j - (1,-5)
e - (3,0)	k - (-1.5,0)
f - (0,0)	l - (0,2.5)

3.

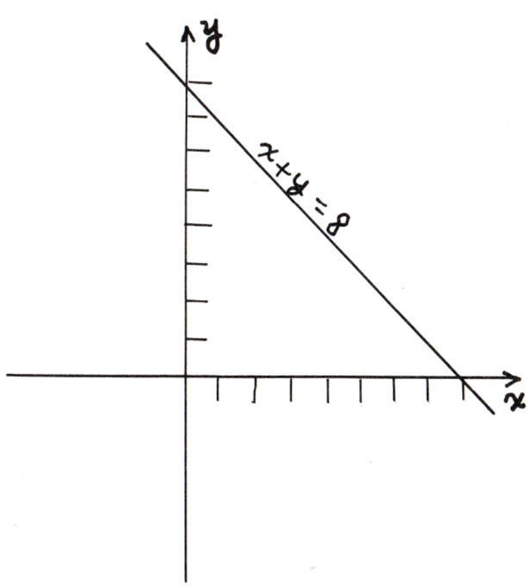

4.

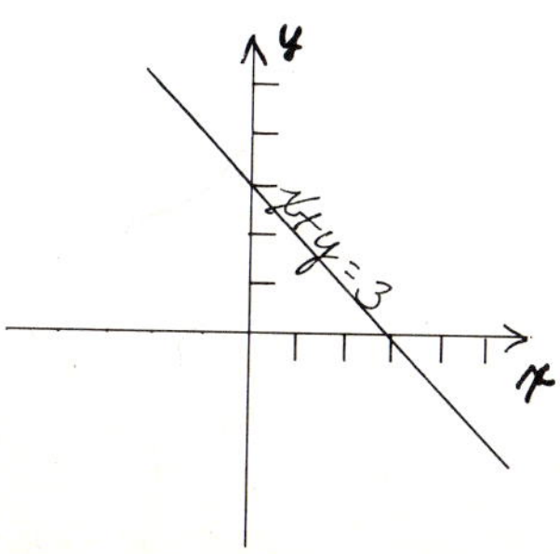

5.

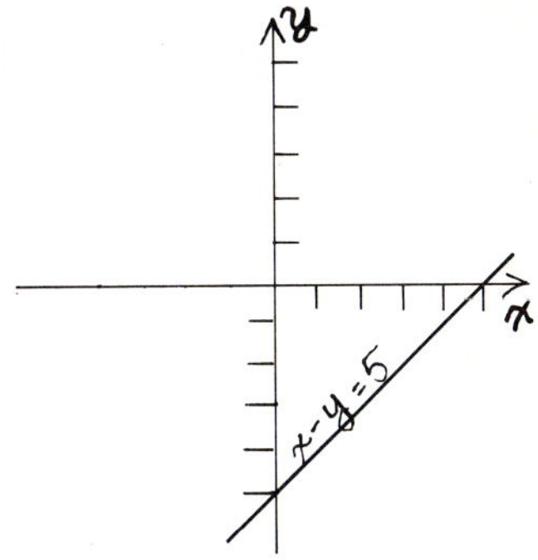

6.

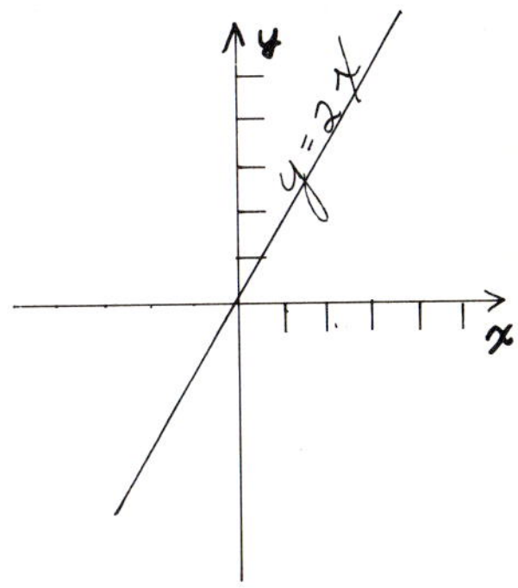

7.

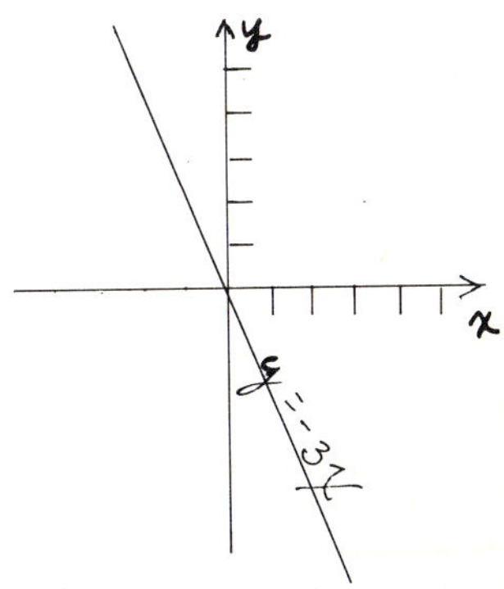

8.

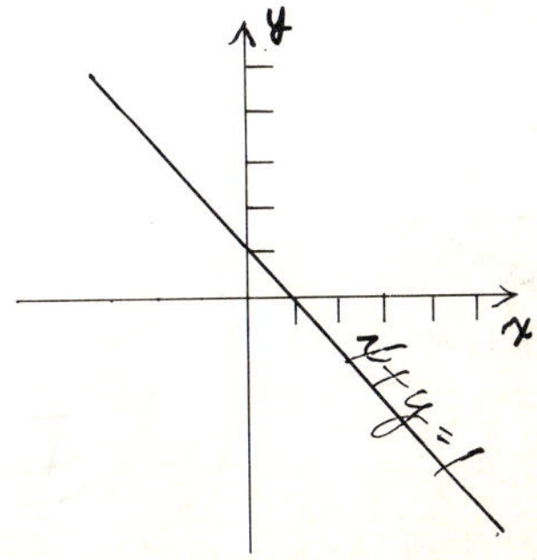

9.

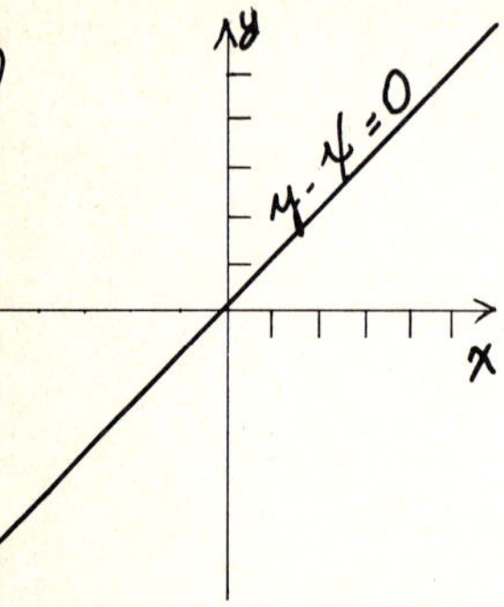

13.

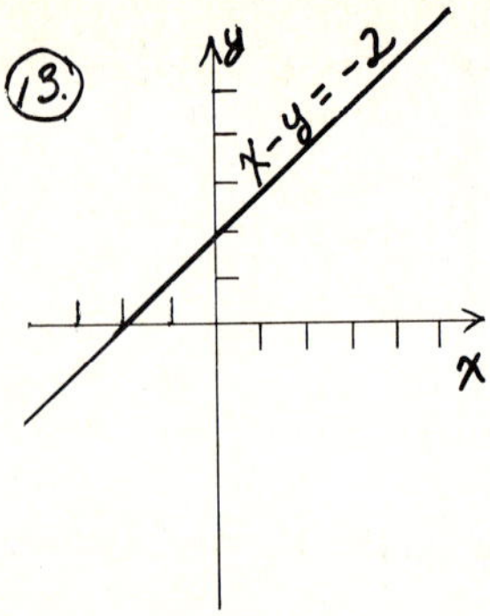

17.

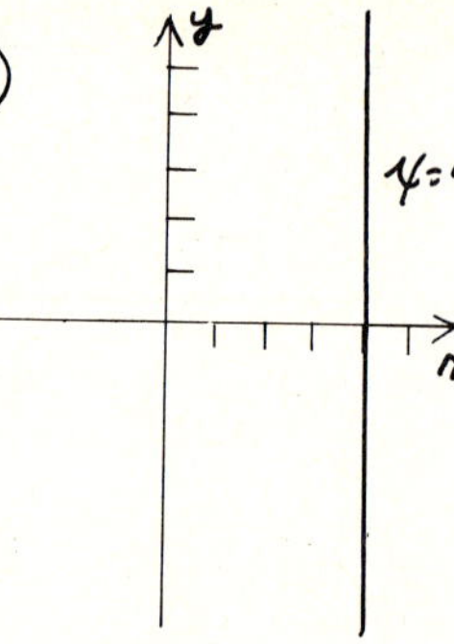

10.

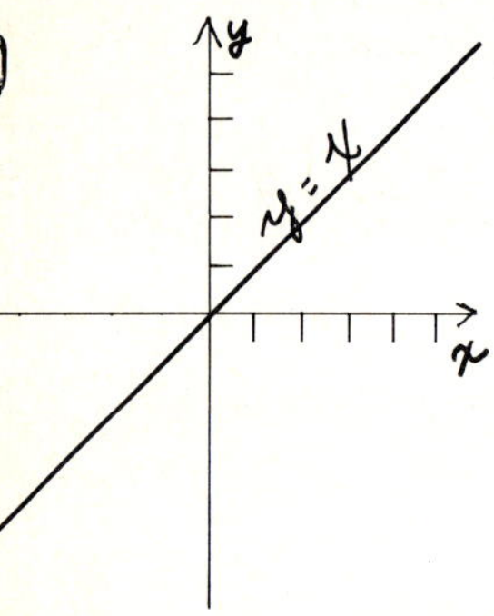

14.

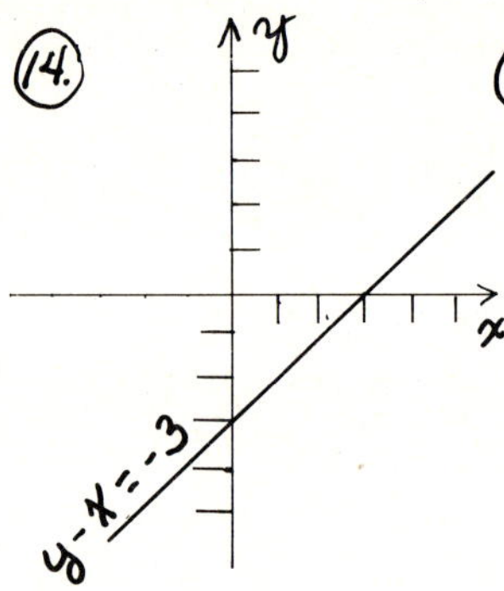

18.

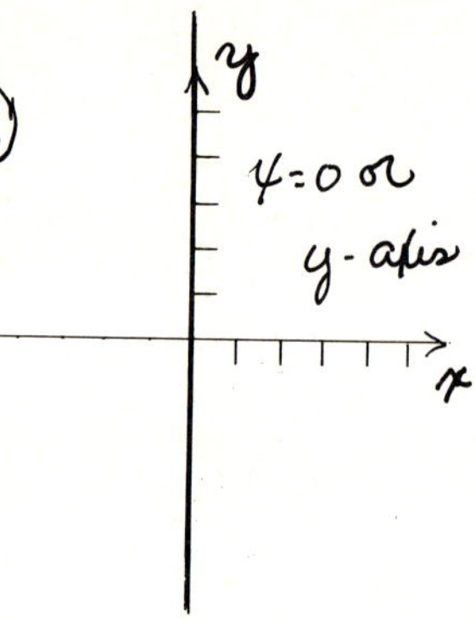

11.

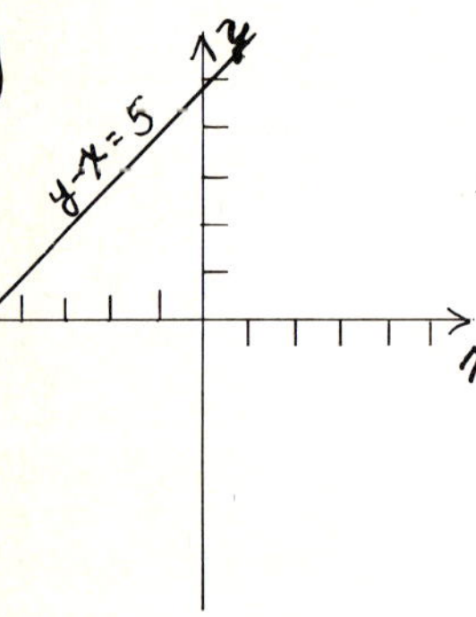

15.

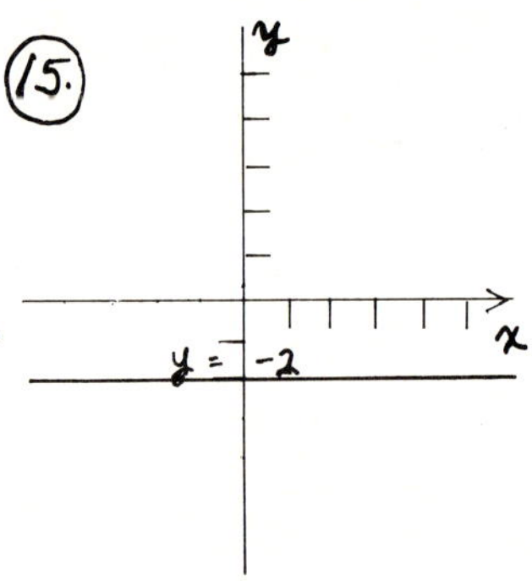

12.

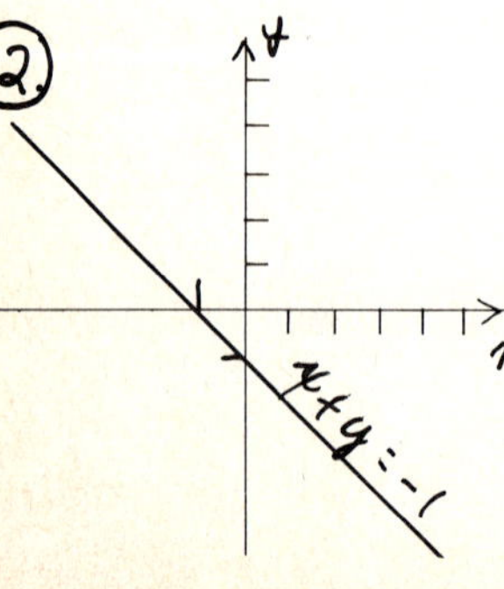

16.

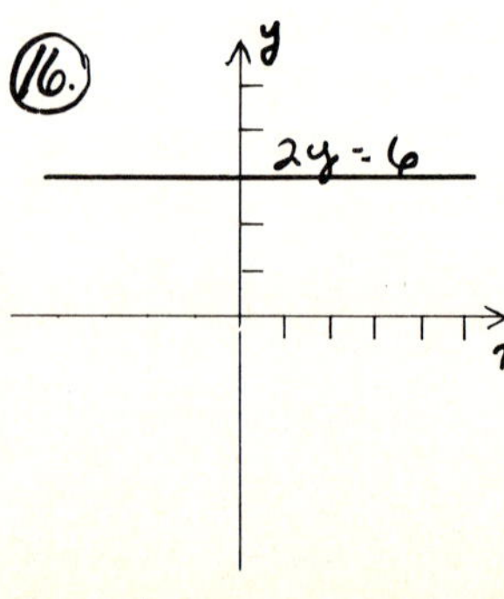

Additional Exercises

1. Graph the following points and identify each point with its x-coordinate and y-coordinate.

(1, 2)	(−4, 0)	($\frac{1}{2}$, 0)
(3, −4)	(2, −1)	(0, −2.5)
(−1, 2)	(3, 4)	
(2, 0)	(4, 3)	
(0, 3)	(−1, −1)	

2. Determine the coordinates of the following points.

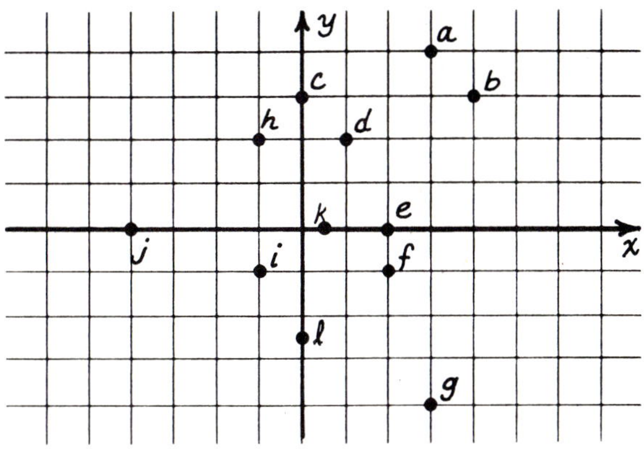

Graph the solution of each of the following equations, using a separate set of axes for each. Write the equation along each line graph.

3. $x + y = 4$
4. $x + y = 2$
5. $x + y = 0$
6. $x + y = -2$
7. $x - y = 5$
8. $x - y = 1$
9. $x - y = -1$
10. $y - x = -1$
11. $y = 3x$
12. $y = -2x$
13. $y = -x$
14. $y = 2x$
15. $y = 1$
16. $y = 0$
17. $2x = 8$
18. $x = -3$

Lesson 4

Advanced Linear Equations in Two Variables

To graph the solutions of the equation $y - 2x = 6$, it is necessary to find a few of the solutions. It is not as easy to "guess" solutions as in the last lesson.

First, we get the equation in a form that will be easier to work with: We get the y on the left side of the equation by itself as follows.

$$\begin{array}{lrcl} & y - 2x & = & 6 \\ \text{Add } 2x \text{ to both sides} & \underline{+\ 2x} & & \underline{+2x} \\ & y + 0 & = & +2x + 6 \\ \text{which is the same as} & y & = & 2x + 6 \end{array}$$

Now, to find a solution of the equation, we choose <u>any number for x,</u> plug it in and find out what y must be in order to make the equation true. That is, we choose a convenient x and then calculate the y for that x. So, for example, <u>if 3 is chosen for x</u> then

$$\begin{aligned} y &= 2x + 6 \\ y &= 2(3) + 6 \\ y &= 6 + 6 \\ y &= 12 \end{aligned}$$

Therefore, one solution of the equation $y - 2x = 6$ is (3, 12). Now w start a table of solutions.

x	y
3	12

To find another solution, choose a different number for x, for example, 2. Plug it in and find out what y must be to make the equation true.

$$y = 2x + 6$$
$$y = 2(2) + 6$$
$$y = 4 + 6$$
$$y = 10$$

Therefore, another solution of the equation $y - 2x = 6$ is (2, 10). The table now has two entries.

x	y
3	12
2	10

More solutions can be found as follows.

If $x = 1$, then $y = 2(1) + 6$ so $y = 8$, and (1, 8) is a solution.
If $x = 0$, then $y = 2(0) + 6$ so $y = 6$, and (0, 6) is a solution.
If $x = -1$, then $y = 2(-1) + 6$ so $y = 4$, and $(-1, 4)$ is a solution.

One way of keeping track of your work is to build a table as follows.

x	$2x + 6$	y	Plot (x, y)
3	$2(3) + 6$	12	(3, 12)
2	$2(2) + 6$	10	(2, 10)
1	$2(1) + 6$	8	(1, 8)
0	$2(0) + 6$	6	(0, 6)
-1	$2(-1) + 6$	4	$(-1, 4)$

These five points can now be plotted on a graph. The solution of $y - 2x = 6$ is the straight line in Figure 1.

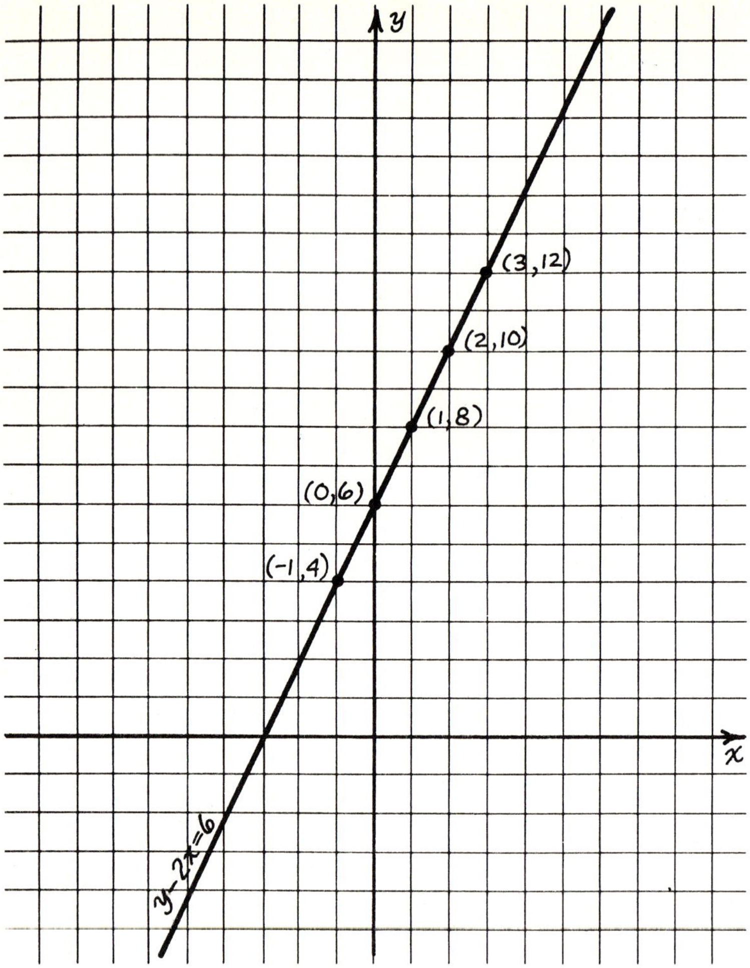

Figure 1

Remember it is not necessary to plot so many points. You only need two, but you can plot three or more points to be sure that they are all on the same line.

EXAMPLE 1 Graph the solutions of the equation $y + 2x = -3$.

First get the y on the left side of the equation by itself as follows.

$$\begin{array}{lrcr} & y + 2x & = & -3 \\ \text{Add } -2x \text{ to both sides} & -2x & & -2x \\ \hline & y & = & -2x - 3 \end{array}$$

x	$-2x - 3$	y	Plot (x, y)
2	$-4 - 3$	-7	$(2, -7)$
1	$-2 - 3$	-5	$(1, -5)$
0	$0 - 3$	-3	$(0, -3)$
-1	$2 - 3$	-1	$(-1, -1)$
-2	$4 - 3$	1	$(-2, 1)$

The graph of $y = -2x - 3$ is shown in Figure 2.

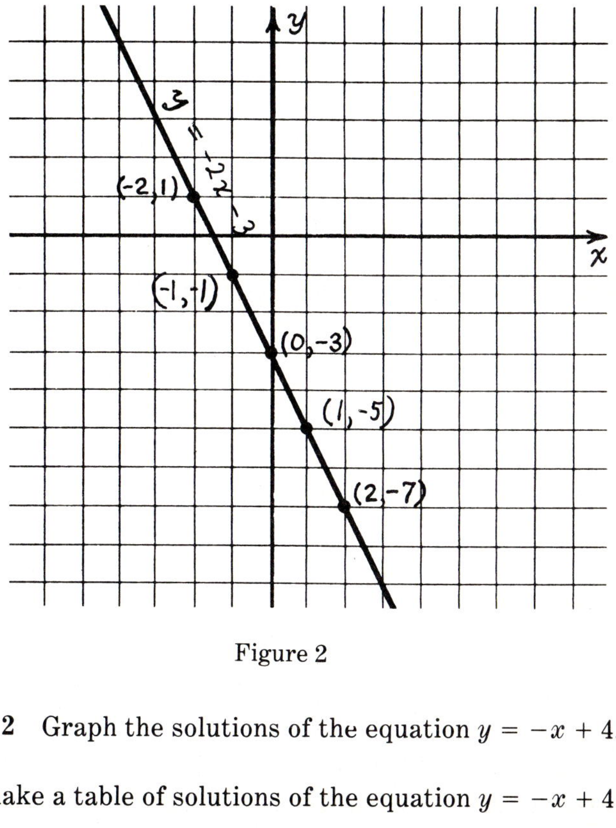

Figure 2

EXAMPLE 2 Graph the solutions of the equation $y = -x + 4$.

Let's make a table of solutions of the equation $y = -x + 4$.

Note: Since $-x$ means $-1(x)$, then $y = -x + 4$ is the same as $y = -1(x) + 4$.

x	$-1x + 4$	y	Plot (x, y)
2	$-1(2) + 4$	2	$(2, 2)$
1	$-1(1) + 4$	3	$(1, 3)$
0	$-1(0) + 4$	4	$(0, 4)$
-1	$-1(-1) + 4$	5	$(-1, 5)$
-2	$-1(-2) + 4$	6	$(-2, 6)$

These five points can now be plotted. The solution of $y = -x + 4$ is the straight line in Figure 3.

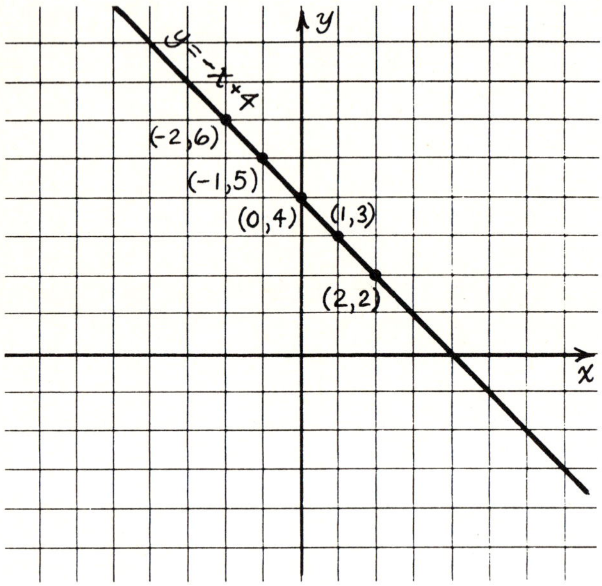

Figure 3

EXAMPLE 3 Graph the solutions of the equation $3y + 3 = x$.

First get the y by itself on the left side of the equation as follows.

Add -3 to both sides

$$\begin{aligned} 3y + 3 &= x \\ -3 &\quad -3 \\ \hline 3y + 0 &= x - 3 \\ 3y &= x - 3 \end{aligned}$$

Divide both sides by 3.

$$\frac{3y}{3} = \frac{x-3}{3}$$

is the same as

$$\frac{3y}{3} = \frac{x}{3} - \frac{3}{3}$$

or

$$y = \frac{x}{3} - 1$$

Since x in this form of the equation is divided by 3, to avoid fractions it is a good idea to pick multiples of 3 for x.

x	$\frac{x}{3} - 1$	y	Plot (x, y)
6	$2 - 1$	1	(6, 1)
3	$1 - 1$	0	(3, 0)
0	$0 - 1$	-1	(0, -1)
-3	$-1 - 1$	-2	(-3, -2)
-6	$-2 - 1$	-3	(-6, -3)

Plotting these solutions we get Figure 4.

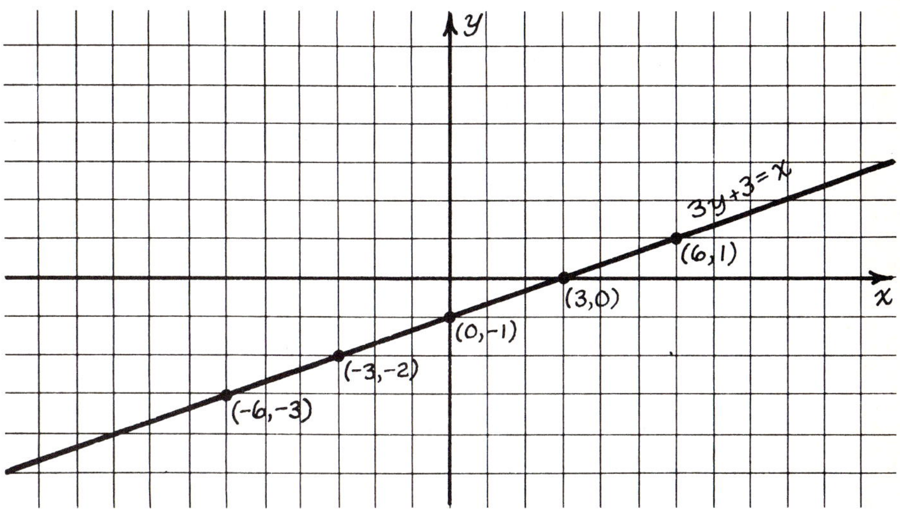

Figure 4

EXAMPLE 4 Graph the solutions of the equation $2y - 5x = 0$.

First get the y by itself on the left side of the equation as follows.

Add $5x$ to both sides

$$\begin{aligned} 2y - 5x &= 0 \\ +5x &\quad 5x \\ \hline 2y &= 5x \end{aligned}$$

Divide both sides by 2.

$$\frac{2y}{2} = \frac{5x}{2}$$

$$y = \frac{5x}{2}$$

Since $5x$ in this form of the equation is divided by 2, it is a good idea to pick multiples of 2 for x to avoid fractions.

x	$\frac{5x}{2}$	y	Plot (x, y)
2	$\frac{5(2)}{2}$	5	(2, 5)
0	$\frac{5(0)}{2}$	0	(0, 0)
−2	$\frac{5(-2)}{2}$	−5	(−2, −5)

These three points can be plotted. The solution of $2y - 5x = 0$ is the straight line in Figure 5.

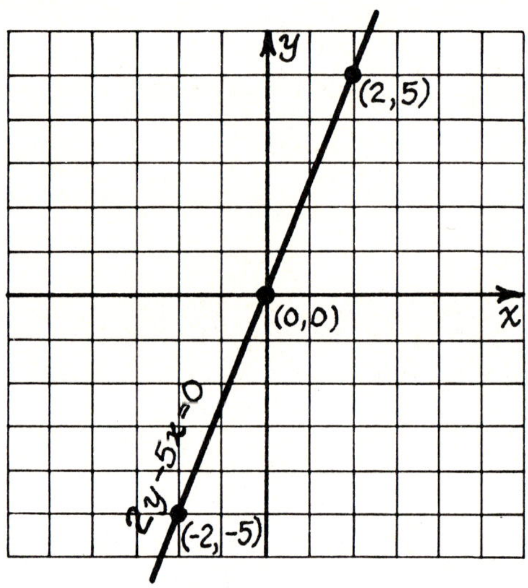

Figure 5

EXAMPLE 5 Graph the solutions of the equation $y - 1.5x = -2.5$.

First get the y by itself on the left side of the equation as follows.

$$\begin{array}{lrcl} & y - 1.5x & = & -2.5 \\ \text{Add } + 1.5x & + 1.5x & & +1.5x \\ \hline & y & = & 1.5x - 2.5 \end{array}$$

x	$1.5x - 2.5$	y	Plot (x, y)
0	$1.5(0) - 2.5$	-2.5	$(0, -2.5)$
1	$1.5(1) - 2.5$	-1	$(1, -1)$
2	$1.5(2) - 2.5$	$+0.5$	$(2, +0.5)$
3	$1.5(3) - 2.5$	$+2$	$(3, 2)$

The solutions of $y - 1.5x = -2.5$ is the straight line graph in Figure 6.

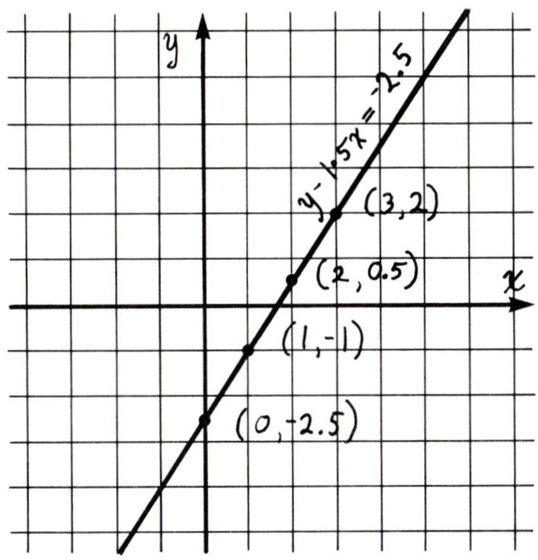

Figure 6

EXAMPLE 6 The line in Figure 7 is the graph of the solutions of an equation. Fill in the missing numbers, so that each of the following is a solution.

a. $(-3, \quad)$ c. $(\quad , +3)$ e. $(0, \quad)$
b. $(\quad , -3)$ d. $(+1, \quad)$ f. $(\quad , 0)$

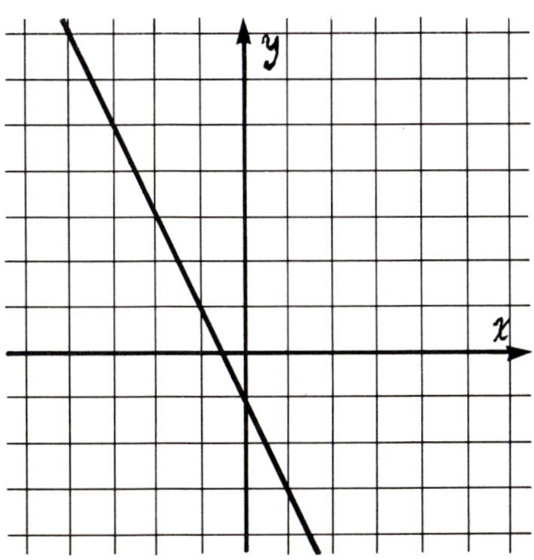

Figure 7

a. For the solution $(-3,\ \ )$, x is -3 and we want to find the missing y. We find -3 on the x-axis and then move in the y-direction (up or down) until we hit the graph. Figure 8 shows that this solution is $(-3, 5)$.

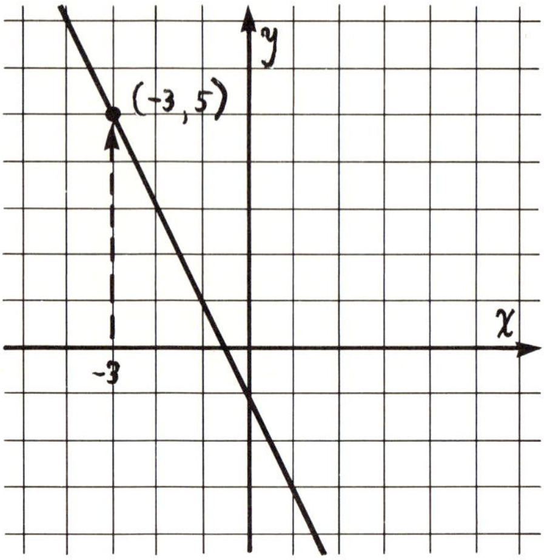

Figure 8

b. For the solution $(\ \ , -3)$, y is -3 and we want to find the missing x. We find -3 on the y-axis and then move in the x-direction (right or left) until we hit the graph. Figure 9 shows that this solution is $(1, -3)$.

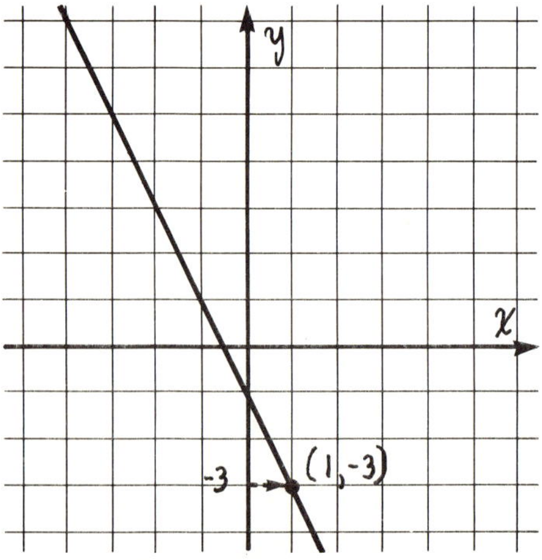

Figure 9

We show the solutions for c, d, e and f in Figure 10.

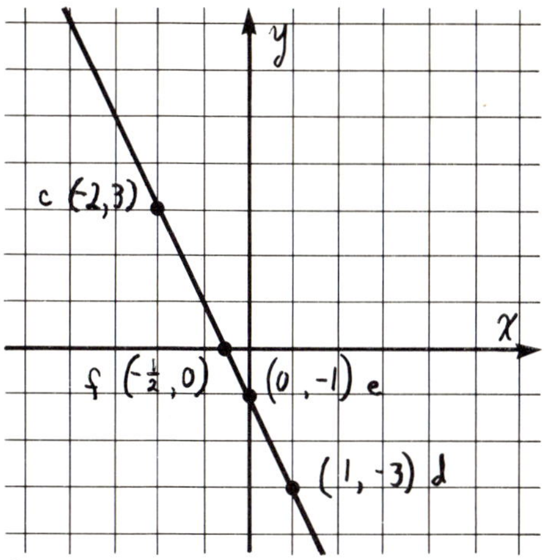

Figure 10

The solutions are:

a. $(-3, 5)$ c. $(-2, +3)$ e. $(0, -1)$
b. $(1, -3)$ d. $(+1, -3)$ f. $(-\frac{1}{2}, 0)$

Exercises

1. For each of the following values of x, find what y must be in order to make the equation, $y = -4x + 7$, true.
   a. x is 5
   b. x is 0
   c. x is -3
   d. x is $\frac{1}{2}$
   e. x is -1.5

Graph the solutions of each of the following equations, using a separate set of axes for each. Write the original equation on each line graphed.

2. $y = x + 1$
3. $y = x + 2$
4. $y = 2x$
5. $y = 2x + 1$
6. $y = 3x - 1$
7. $y = 3x - 2$
8. $y = -x + 3$
9. $y = -x - 1$
10. $y = x$
11. $y = -x$
12. $y = -3$
13. $y = 4$
14. $x = 2$
15. $x = -1$
16. $y = \frac{1}{2}x + 2$
17. $y = \frac{1}{3}x - 1$
18. $y = 1.5x + 1$
19. $y = 2.5x - 2$
20. $y - 2x = 3$
21. $x + y = 3$
22. $x - y = 3$
23. $2x + y = 1$
24. $2x + 3y = 6$
25. $2y - 3x = 4$
26. $3x - 2y = 2$
27. $3x - 4y = 4$
28. $3y - 4x = 0$
29. $5y + 0x = 10$
30. $4y = -8$
31. $2y = 3x$
32. $3y = -2x$

33. The line below is the graph of the solutions of an equation. Fill in the missing numbers, so that each of the following is a solution.

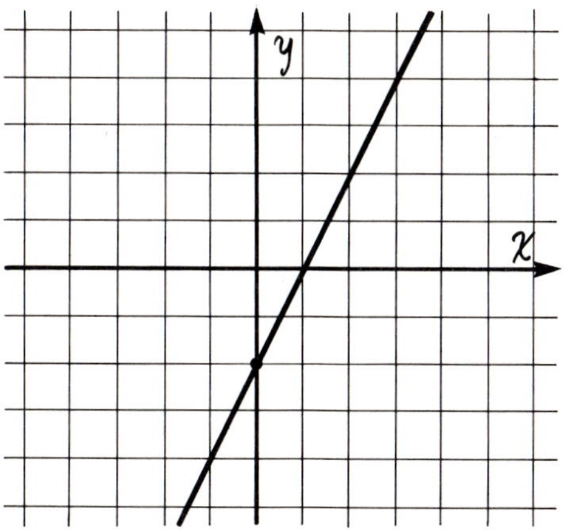

a. (+1,   )   c. (, 2)   e. (, 4)

b. (0,   )   d. (, −4)   f. ($\frac{1}{2}$,   )

Answers to Exercises

(1.) $y = -4x + 7$

x	$-4x+7$	y
5	$-4(5)+7$	-13
0	$-4(0)+7$	7
-3	$-4(-3)+7$	19
$\frac{1}{2}$	$-4(\frac{1}{2})+7$	5
-1.5	$-4(-1.5)+7$	13

(2.)

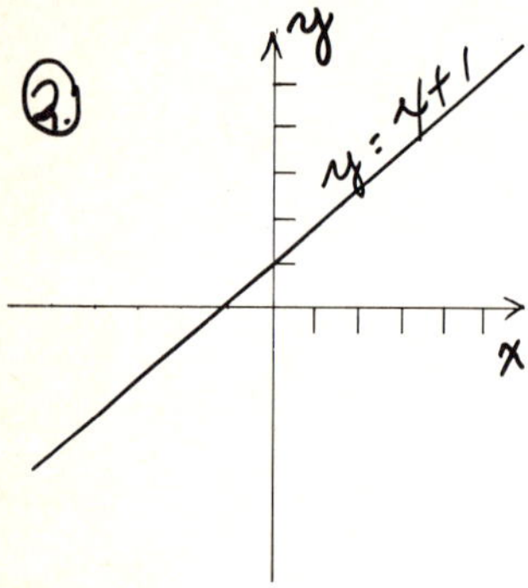

(3.)

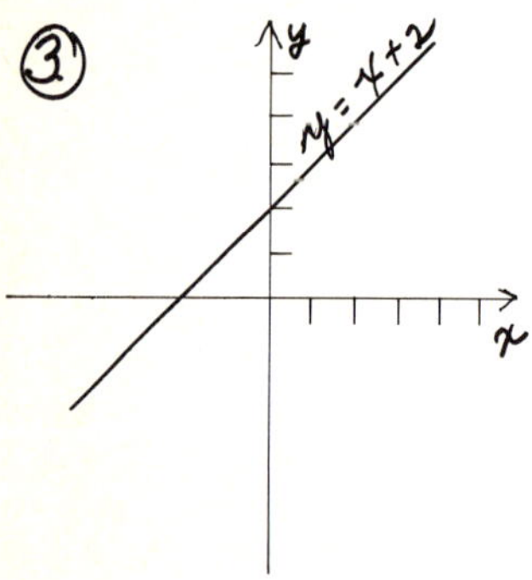

(4.)

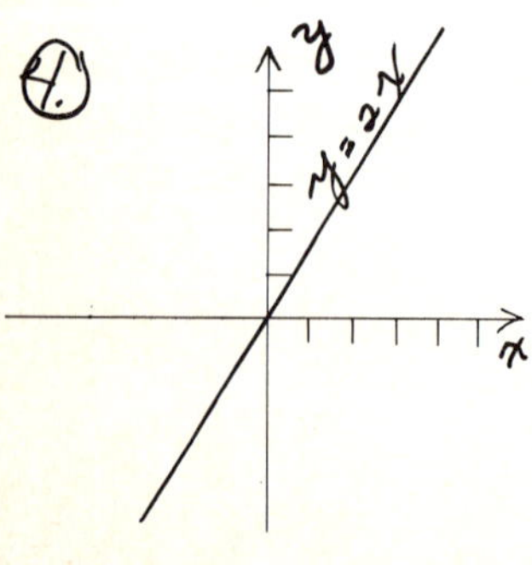

(5.)

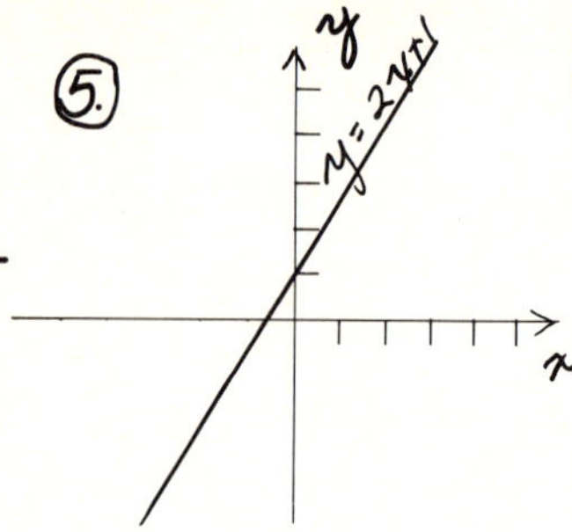

(6.)

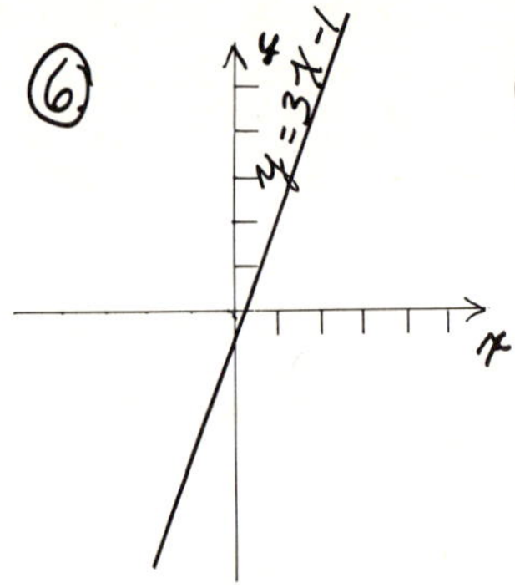

(7.)

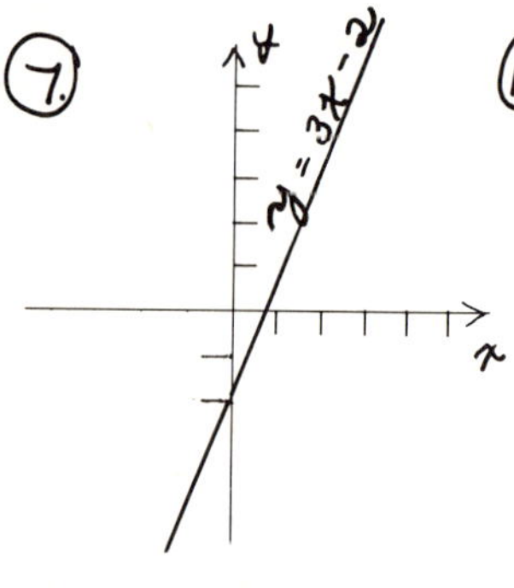

(8.)

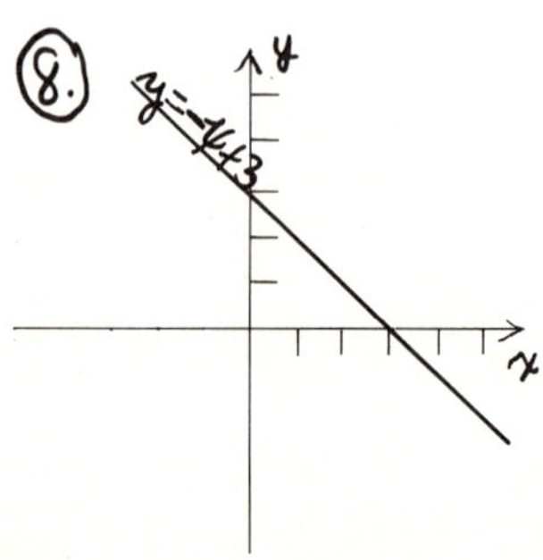

(9.)

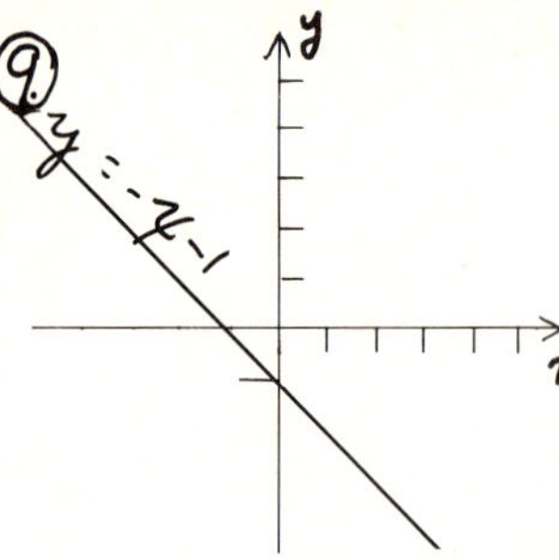

(10.)

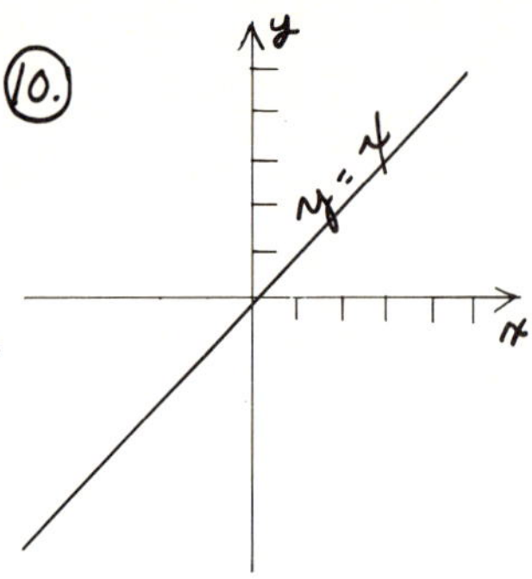

(11.)

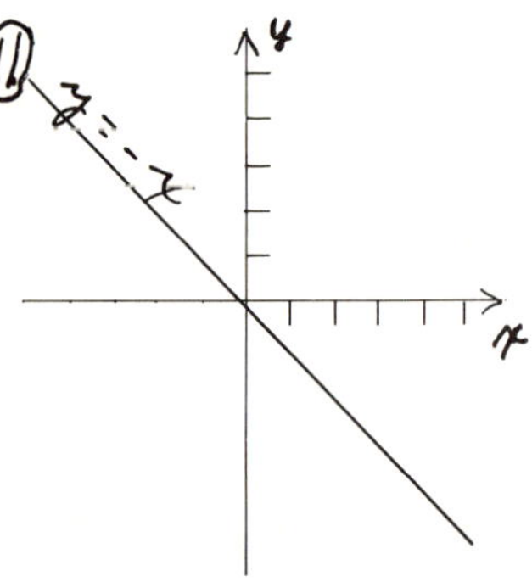

(12.)

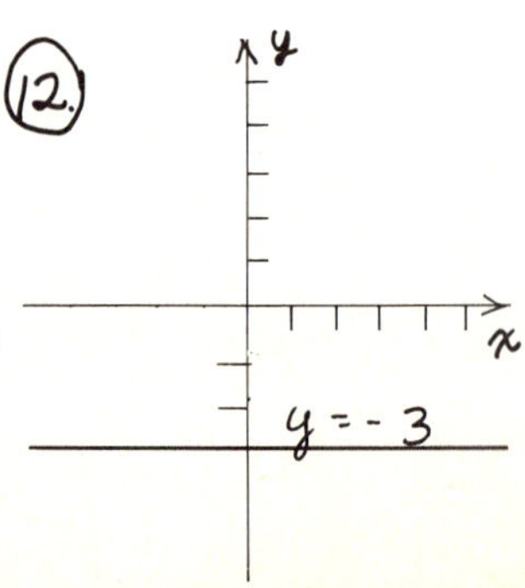

13.

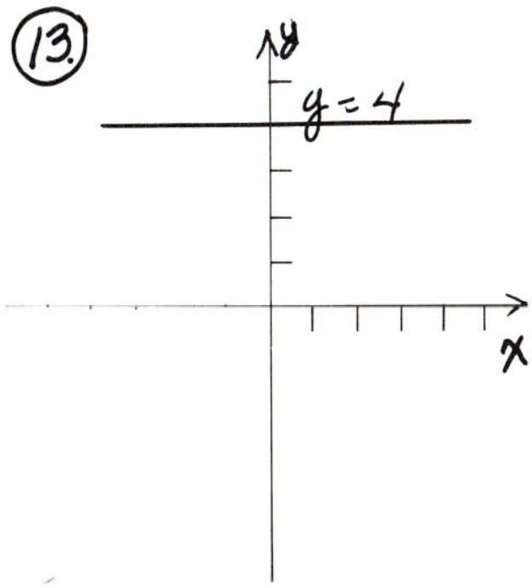

14.

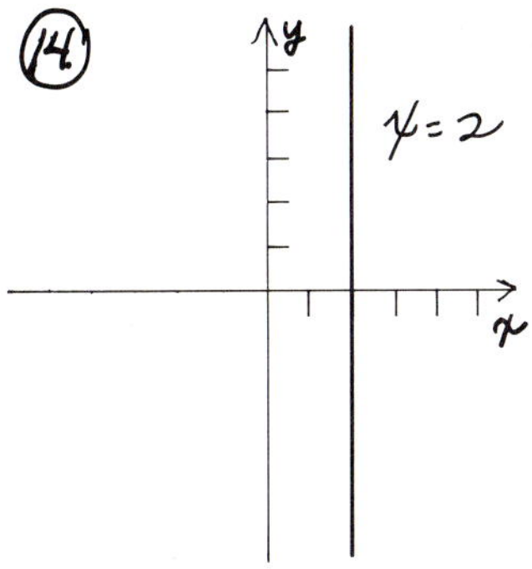

15.

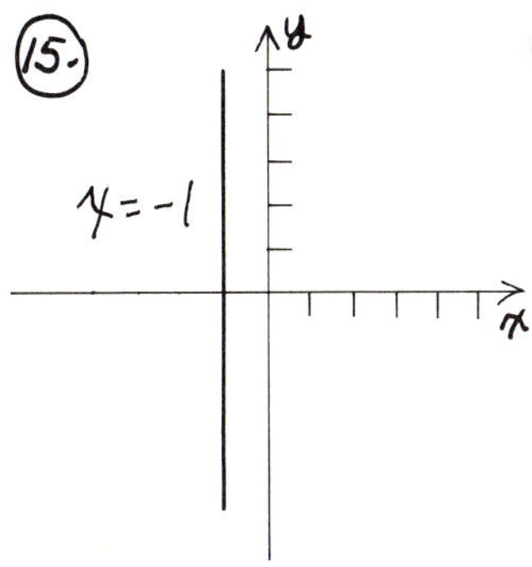

16.

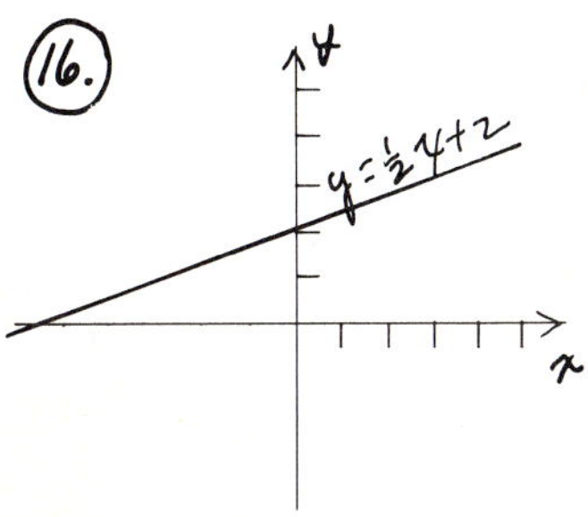

17.

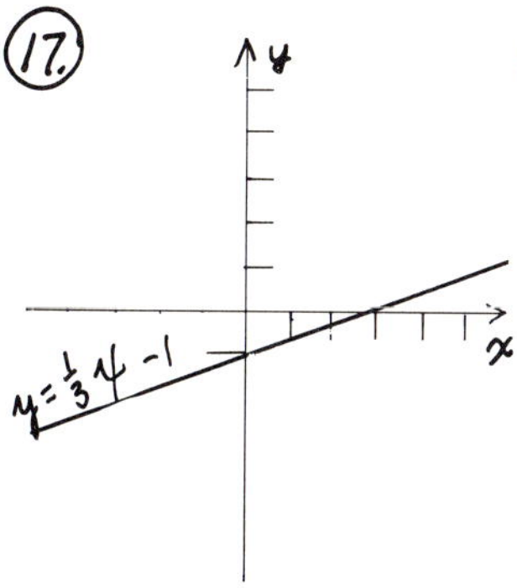

18.

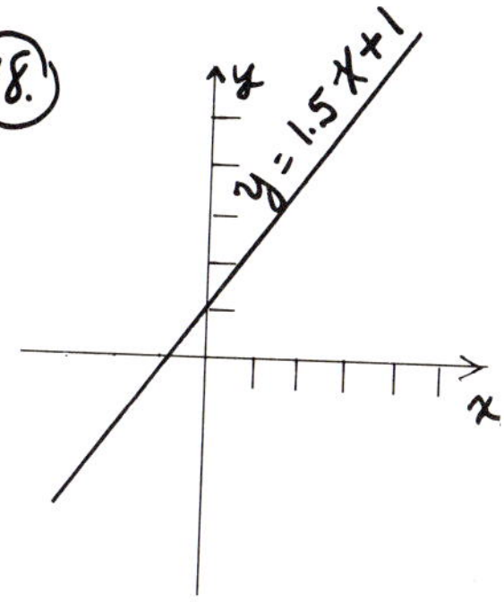

19.

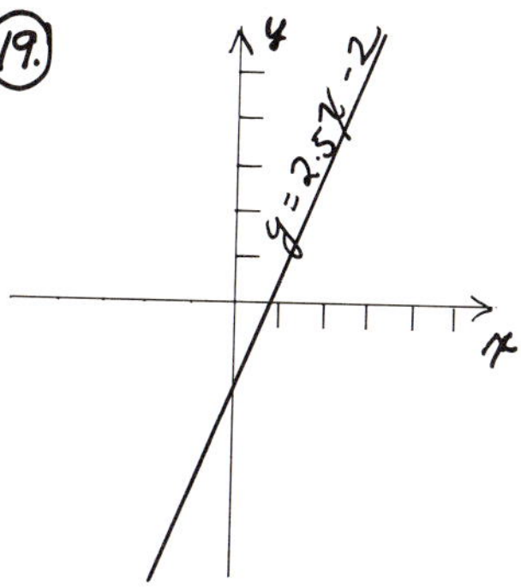

20.

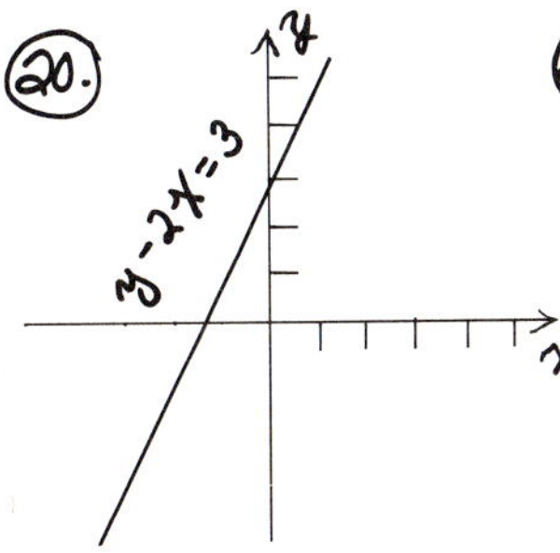

21.

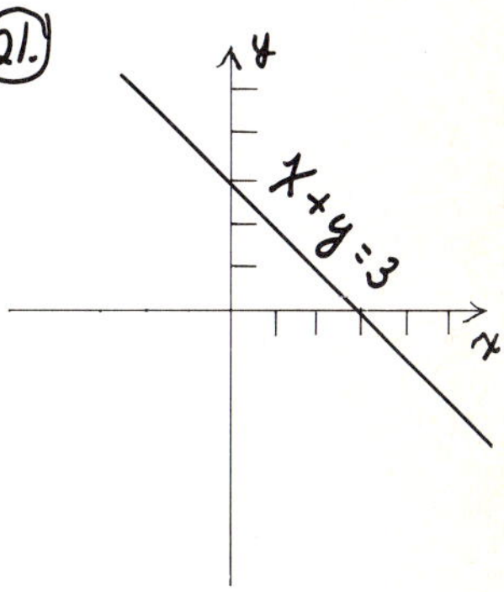

22.

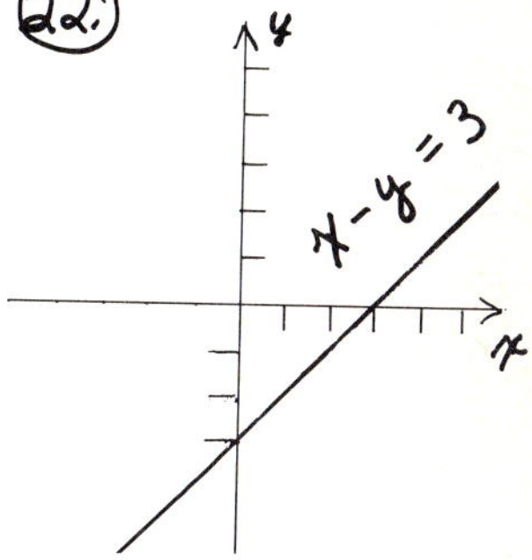

23.

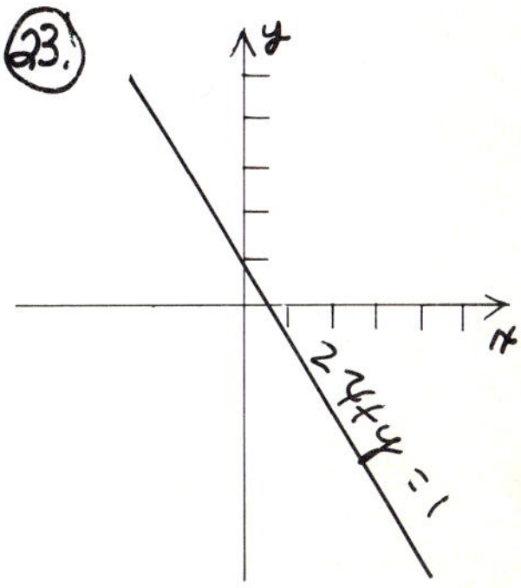

24.

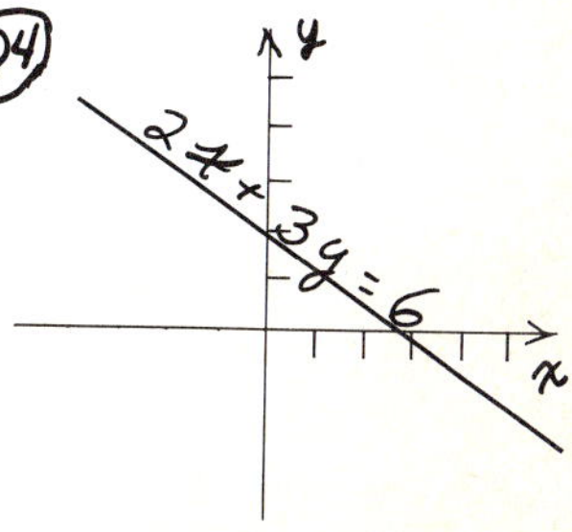

25.

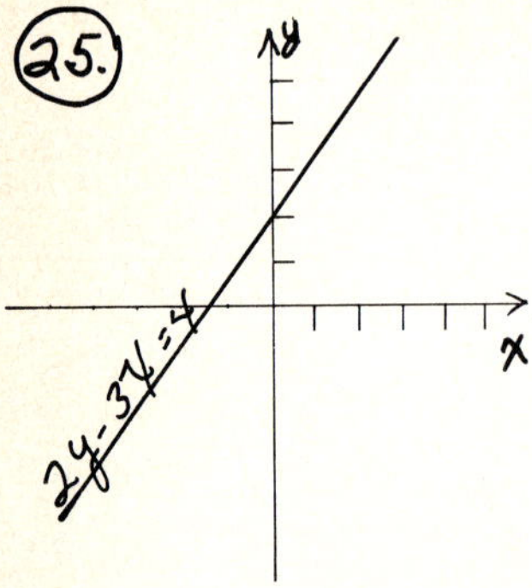

29.

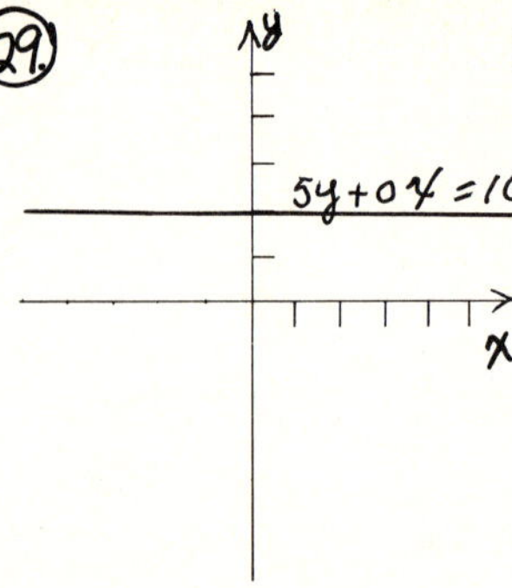

33.

a. (1,0)
b. (0,-2)
c. (2,2)
d. (-1,-4)
e. (3,4)
f. ($\frac{1}{2}$,-1)

26.

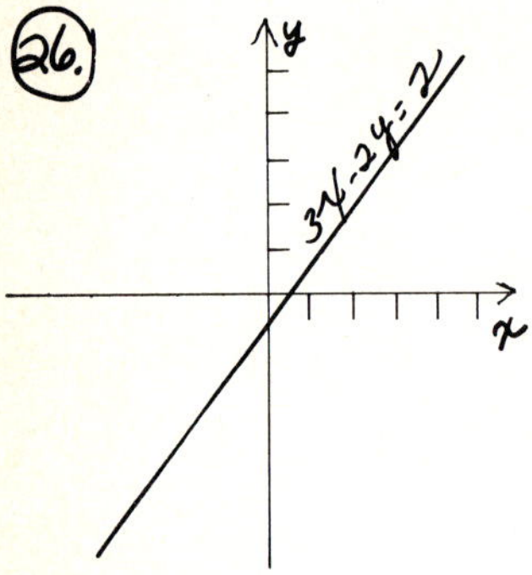

30.

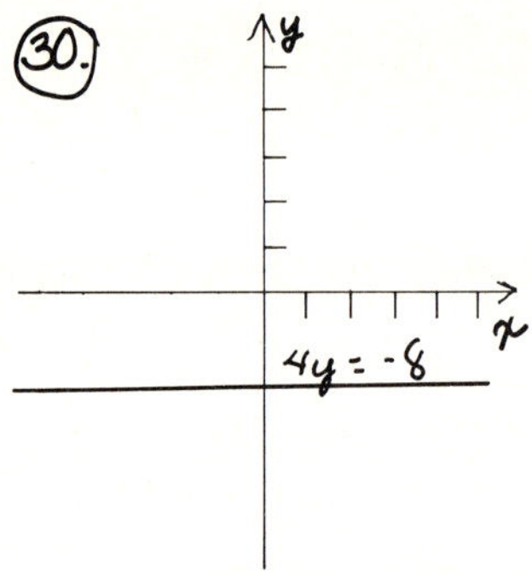

27.

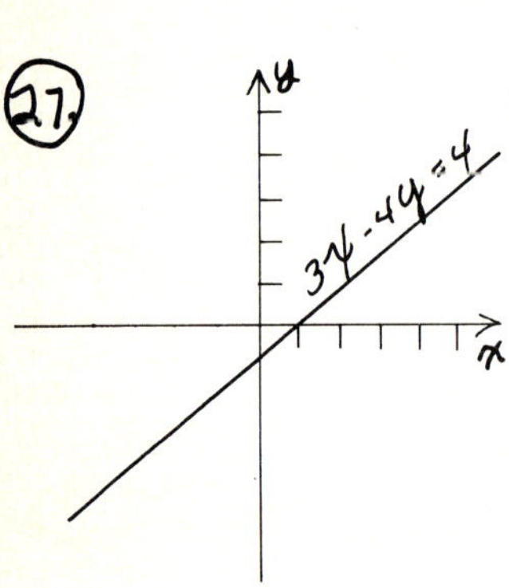

31.

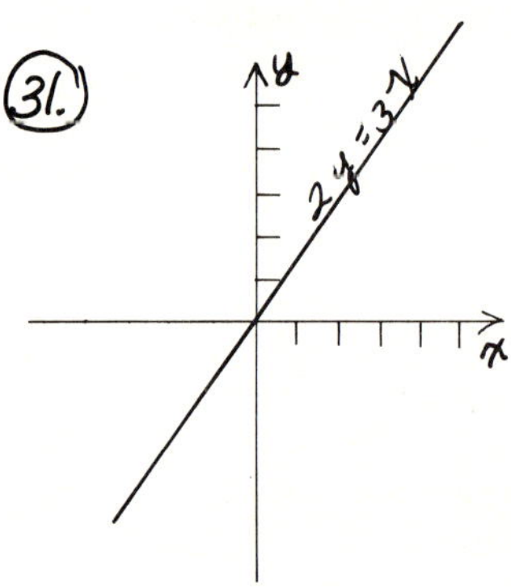

28.

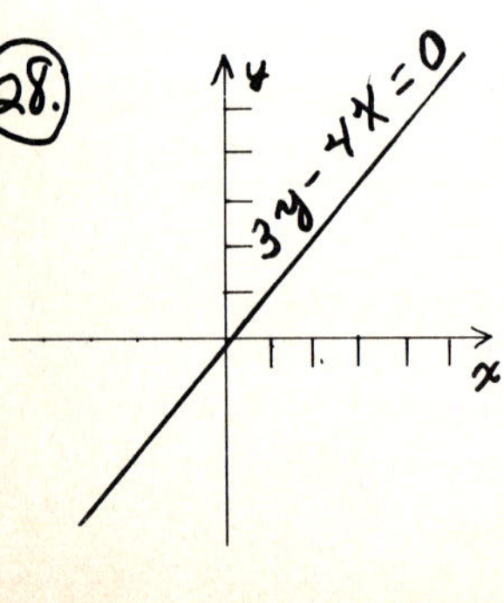

32.

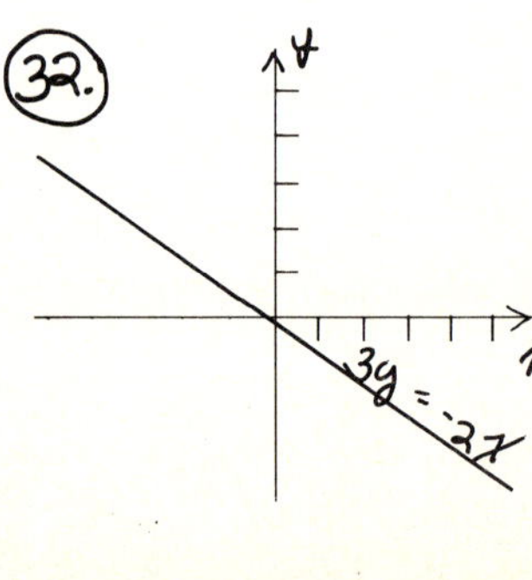

Additional Exercises

1. For each of the following values of x, find what y must be in order to make the equation $y = -3x + 4$ true.
   a. x is 4
   b. x is 0
   c. x is -2
   d. x is $-\frac{1}{3}$
   e. x is -2.5

Graph the solutions of each of the following equations, using a separate set of axes for each. Write the original equation on each line graphed.

2. $y = x + 3$
3. $y = x + 4$
4. $y = 3x$
5. $y = 3x + 1$
6. $y = 2x - 1$
7. $y = 2x - 3$
8. $y = -x + 2$
9. $y = -x - 3$
10. $2y = 2x$
11. $3y = -3x$
12. $y = -4$
13. $y = 2$
14. $x = 5$
15. $x = -3$
16. $y = \frac{1}{3}x + 1$
17. $y = \frac{1}{4}x - 2$
18. $y = 3.5x - 2$
19. $y = -1.5x + 2$
20. $y - 3x = 1$
21. $y - 2x = 4$
22. $x + y = 1$
23. $x - y = 1$
24. $3x + y = 2$
25. $3x + 4y = 12$
26. $4y - 5x = 8$
27. $2x - 3y = 3$
28. $4x - 3y = 3$
29. $5y - 2x = 0$
30. $3y - x = 0$

31. $3y = -9$
32. $2y = 5x$
33. $5y = -2x$

34. The line below is the graph of the solutions of an equation. Fill in the missing numbers, so that each of the following is a solution.

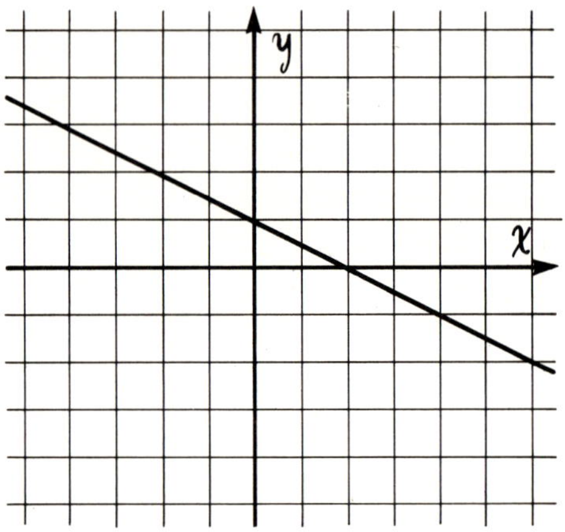

a. (2,   ) c. (, −1) e. (, 2)
b. (0,   ) d. (−1,   ) f. (−4,   )

Lesson 5

Slope

In this lesson we build a way of measuring the steepness of a line. **Slope** is the word we use to describe steepness.

Man A is walking up a steeper line than man B. Since Line A is steeper than Line B, we say Line A has a greater slope than Line B. We also say that the slope is positive if the line goes up as you walk from left to right. Lines A and B have positive slope. If the line goes down as you walk from left to right, the slope of the line is negative. So Man C is on a line with negative slope. Be careful! You must always think "left to right."

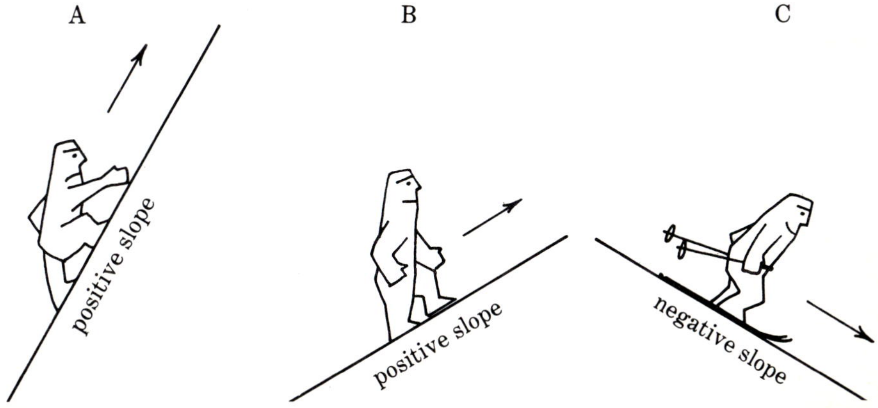

To measure the slope of the line in Figure 1 notice that as the man goes from P to Q he goes up 3 units while going to the right one unit. We say the slope of this line is $\frac{3}{1}$ which is +3.

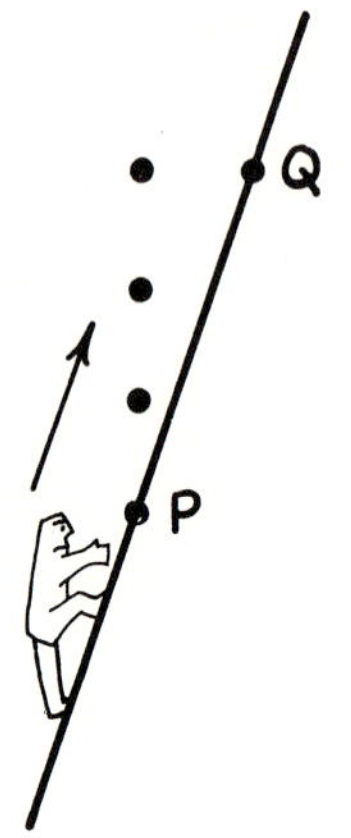

Figure 1

To measure the slope of the line in Figure 2 notice that as the man goes from A to B he goes up 2 units while going to the right one unit. The slope is $\frac{2}{1}$ which is +2. Suppose the man walks from A to C. Then he goes up 4 units while going 2 units to the right. The slope of the line is $\frac{4}{2}$ which is +2. Notice we get the same slope no matter which two points on the line we pick.

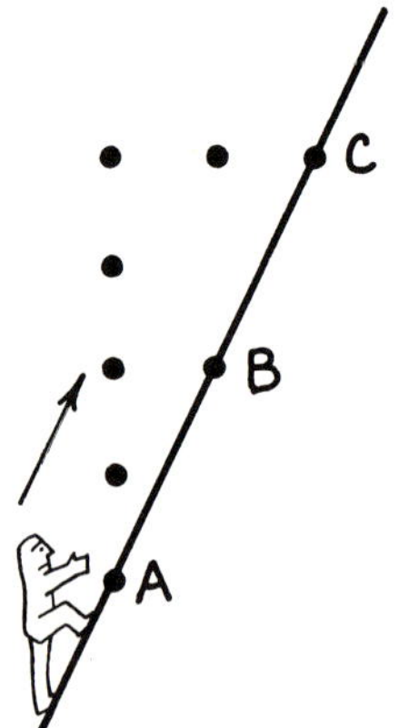

Figure 2

Let's measure the slope of the line in Figure 3 in three different ways.

(1) From P to R. Two units up, two units to the right. Slope is $\frac{2}{2}$ which is 1.
(2) From P to S. Three units up, three units to the right. Slope is $\frac{3}{3}$ or 1.
(3) From Q to R. One unit up, one unit to the right. Slope is $\frac{1}{1}$ or 1.

Note again that no matter what two points on this line we pick, we get the same slope, +1.

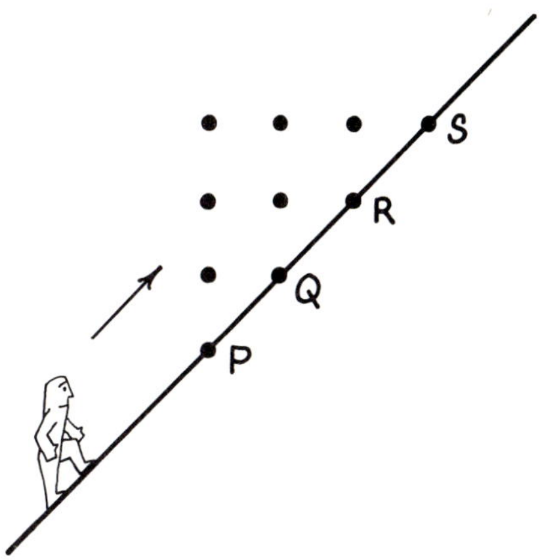

Figure 3

Measure the slope of the line in Figure 4. From D to E we go up two units and to the right three units. The slope is $+\frac{2}{3}$.

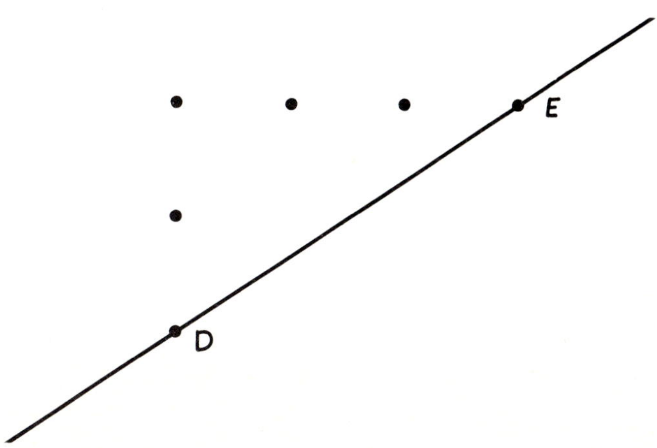

Figure 4

To measure the slope of the line from P to Q in Figure 5, we go down five units and to the right two units. So the slope is $\frac{-5}{2}$. Notice that since we went down five units, we use negative five. Since $\frac{-5}{2}$ is the same as $-\frac{5}{2}$, the slope of this line is a negative number. Just as in graphing, we use positive numbers for $\overset{\uparrow}{\text{Up}}$ and $\overset{\rightarrow}{\text{Right}}$ and negative numbers for $\overset{\downarrow}{\text{Down}}$ and $\overset{\leftarrow}{\text{Left}}$.

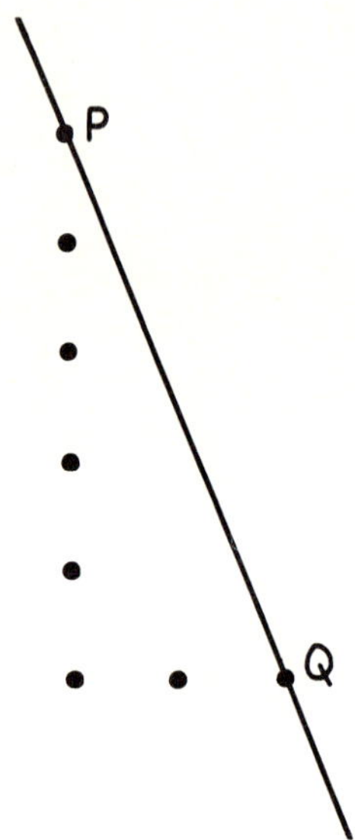

Figure 5

For each of the lines in Figures 1 thru 5 the slope is the change in height divided by the change in level distance.

$$\text{Slope} = \frac{\text{Change in height}}{\text{Change in level distance}}$$

Shorthand for the words "change in height" is Δy (pronounced delta y). Shorthand for the words "change in level distance" is Δx (pronounced delta x).

$$\text{Slope} = \frac{\Delta y}{\Delta x}$$

EXAMPLE 1 Find the slope of the line in Figure 6.

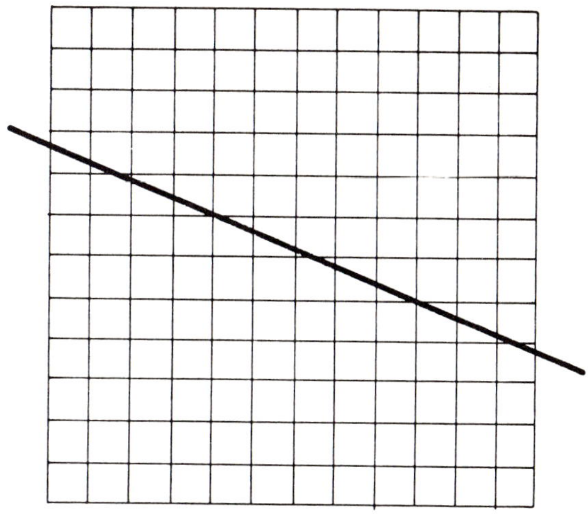

Figure 6

First, select any two points on the line. In Figure 7, we selected points A and B.

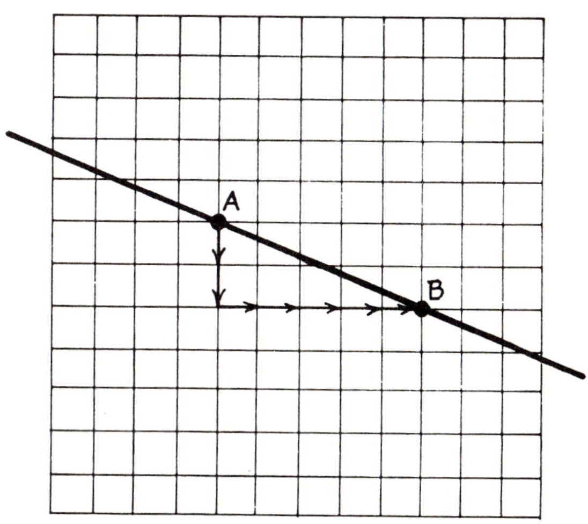

Figure 7

From A to B we go 2 units down (−2) and 5 units to the right (+5). Thus $\Delta y = -2$; $\Delta x = +5$.

So the slope of the line $= \dfrac{\Delta y}{\Delta x} = \dfrac{-2}{+5} = -\dfrac{2}{5}$.

EXAMPLE 2 Find the slope of the line in Figure 8.

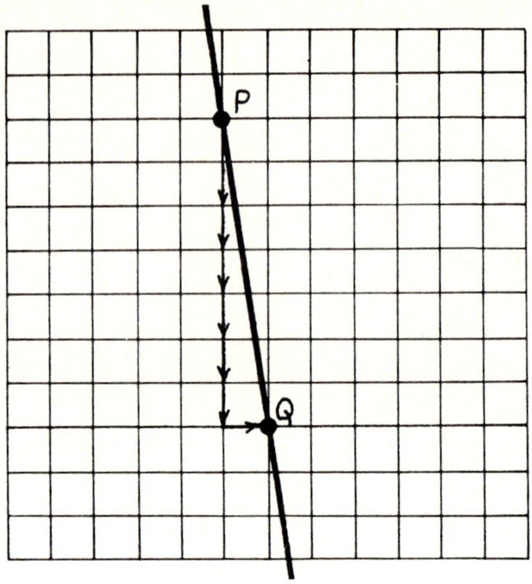

Figure 8

To find the slope of the line in Figure 8, select two points, say P and Q. From P to Q, we go 7 units down (-7) and 1 unit to the right ($+1$). Thus $\Delta y = -7$ and $\Delta x = +1$.

So the slope of the line $= \dfrac{\Delta y}{\Delta x} = \dfrac{-7}{+1} = -7.$

EXAMPLE 3 Find the slope of the line in Figure 9.

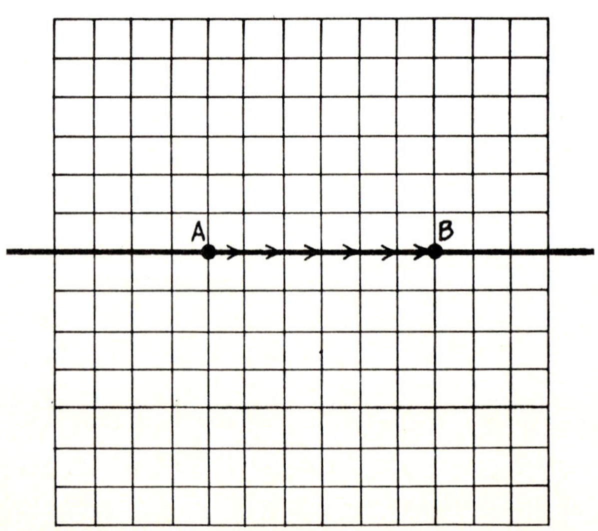

Figure 9

To find the slope of the line in Figure 9, select two points on the line, say A and B. Since A and B are at the same height, Δy is zero (you don't go up or down). Δx is 6. So the slope $= \dfrac{\Delta y}{\Delta x} = \dfrac{0}{6} = 0$.

The slope of every horizontal line is zero, and every line whose slope is zero is horizontal.

EXAMPLE 4 Find the slope of the line in Figure 10.

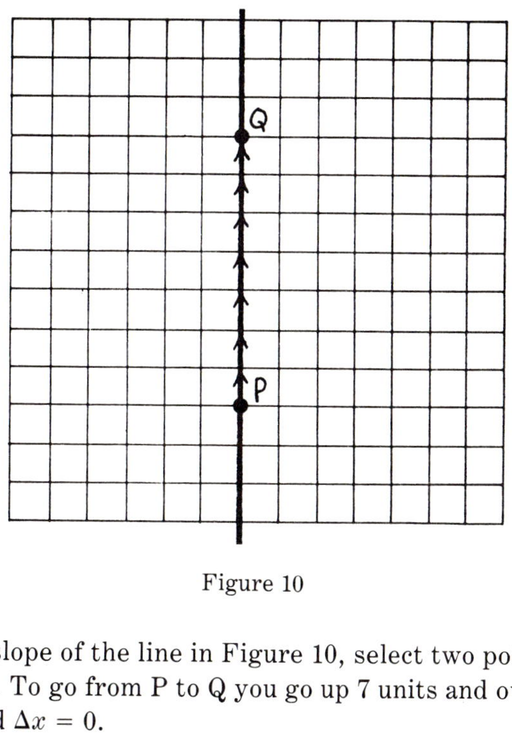

Figure 10

To find the slope of the line in Figure 10, select two points on the line, say P and Q. To go from P to Q you go up 7 units and over 0 units. So $\Delta y = +7$ and $\Delta x = 0$.

But slope $= \dfrac{\Delta y}{\Delta x} = \dfrac{+7}{0}$. Since $\Delta x = 0$ and division by zero has no meaning, we say that the slope of this line is **undefined**; that is, the line doesn't have any slope. The slope of every vertical line is undefined, and every line with undefined slope (that is, without a slope) is vertical.

In all the examples above we were given a line, and the job was to find the slope. Now we will be given a slope, and the job will be to draw a line with that slope.

EXAMPLE 5 Draw a line with a slope of $\frac{5}{7}$.

To draw a line with a slope of $\frac{5}{7}$ you write

$$\text{Slope} = \frac{5}{7} = \frac{\Delta y}{\Delta x}$$

So we can let $\Delta y = +5$ and $\Delta x = +7$.

Start from any point in Figure 11, say P. Go up 5 units (since $\Delta y = +5$), go to the right 7 units (since $\Delta x = +7$), and make a dot; call it Q. Draw a line through P and Q. This line has the slope of $\frac{5}{7}$.

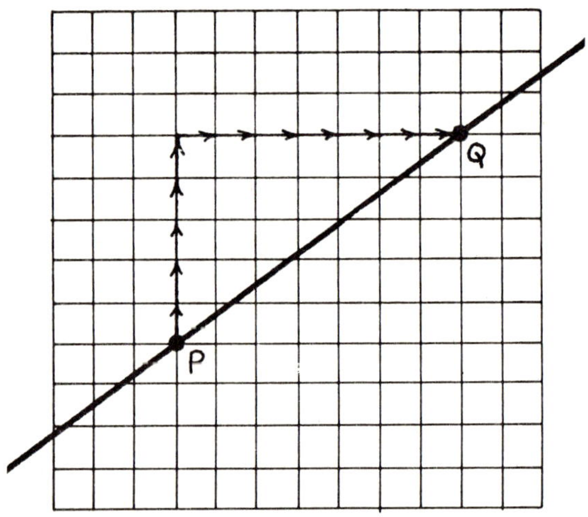

Figure 11

Note: The line in Figure 11 goes up as we go from left to right. This is what we expected since its slope, $\frac{5}{7}$, is positive.

EXAMPLE 6 Draw a line with slope $-\frac{2}{3}$.

To draw a line with slope $-\frac{2}{3}$ (since $-\frac{2}{3} = \frac{-2}{3}$), we write

$$\text{Slope} = \frac{-2}{3} = \frac{\Delta y}{\Delta x}$$

So we can let $\Delta y = -2$ and $\Delta x = +3$.

Start with any point in Figure 12, say A. Go down two units (since $\Delta y = -2$), go 3 units to the right (since $\Delta x = +3$), make a dot; call it B. To be sure we did not make any mistakes we can repeat the process starting from B, and obtain another point C. Draw a straight line through A, B and C. This line has the slope of $-\frac{2}{3}$.

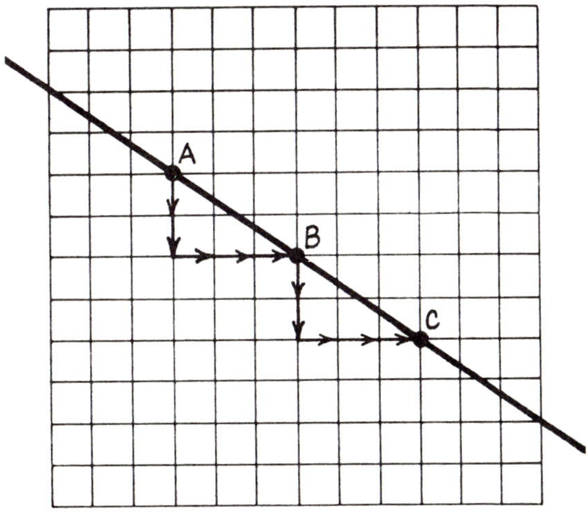

Figure 12

Note: The line in Figure 12 goes down as we go from left to right. This is what we expected since its slope, $-\frac{2}{3}$, is negative.

EXAMPLE 7 Draw a line with slope 3.

To draw a line with slope 3, we write

$$\text{Slope} = 3 = \frac{\Delta y}{\Delta x}$$

It would be nice to write 3 as a fraction, say $\frac{3}{1}$ or $\frac{6}{2}$ or $\frac{9}{3}$ or something else equal to 3. Let's choose $\frac{3}{1}$.

$$\text{Slope} = \frac{3}{1} = \frac{\Delta y}{\Delta x}$$

So we can let $\Delta y = +3$ and $\Delta x = +1$.

Start with any point in Figure 13, say P. Go up 3 units (since $\Delta y = +3$), go 1 unit to the right (since $\Delta x = +1$), and make a dot; call it Q. Draw a line through P and Q. This line has the slope of 3.

Suppose we had chosen $3 = \frac{6}{2}$.

$$\text{Slope} = \frac{6}{2} = \frac{\Delta y}{\Delta x}$$

And again, starting at P, go up 6 units (since $\Delta y = +6$), go 2 units to the right (since $\Delta x = +2$), and make a dot; call it R. Notice that the point R is on the line that we drew through P and Q.

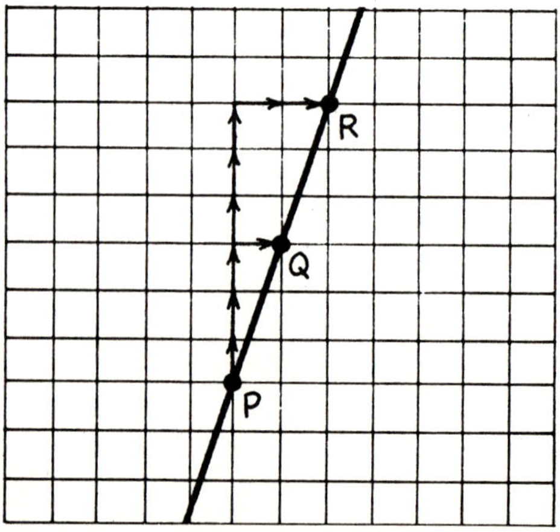

Figure 13

EXAMPLE 8 Draw a line with slope 0.6.

To draw a line with slope 0.6, we change the decimal number to a fraction. 0.6 can be written as $\frac{6}{10}$ or $\frac{3}{5}$.

$$\text{Slope} = 0.6 = \frac{3}{5} = \frac{\Delta y}{\Delta x}$$

So we can let $\Delta y = +3$ and $\Delta x = +5$.

Start with any point in Figure 14, say P. Go up 3 units and then go 5 units to the right and make a dot. We labeled this dot Q. A line through P and Q has slope 0.6.

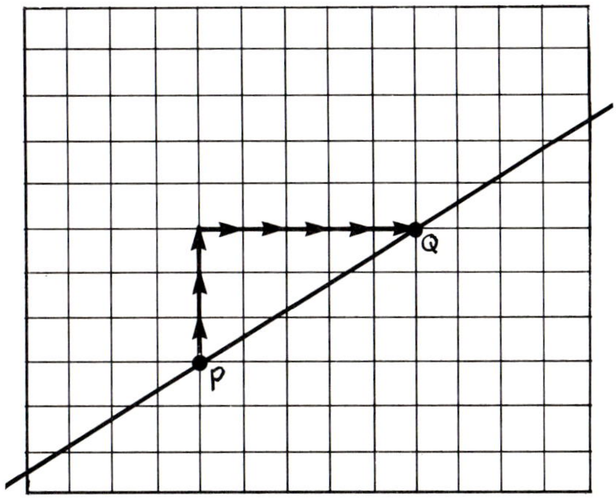

Figure 14

Exercises

1. Calculate the slope of the line in Figures 15 and 16.

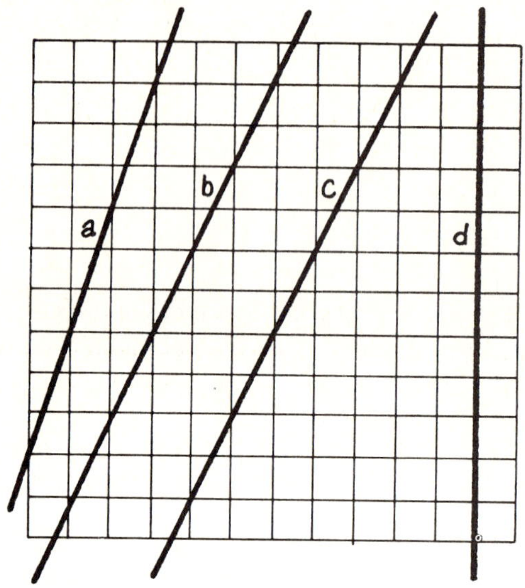

Figure 15

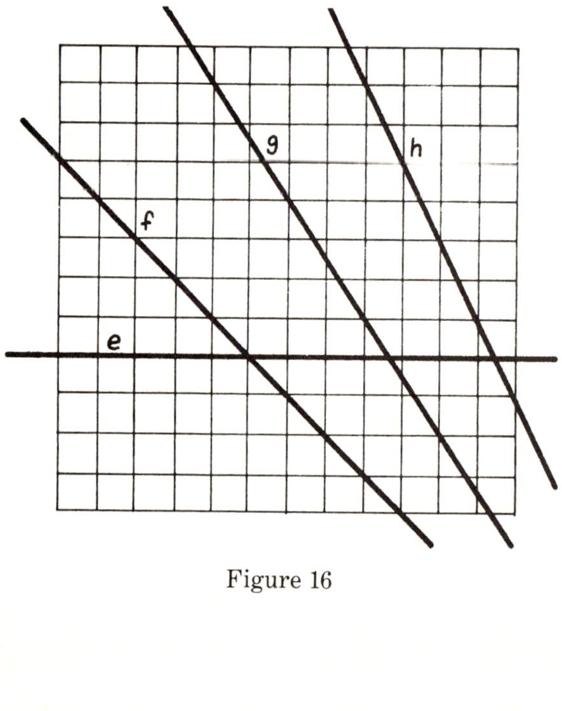

Figure 16

2. Match each line in Figures 17 and 18 with one of the given slopes.

<u>Slope</u>

$\frac{2}{3}$

-3

0

$\frac{3}{4}$

undefined

-1

$-\frac{2}{5}$

1

$\frac{1}{2}$

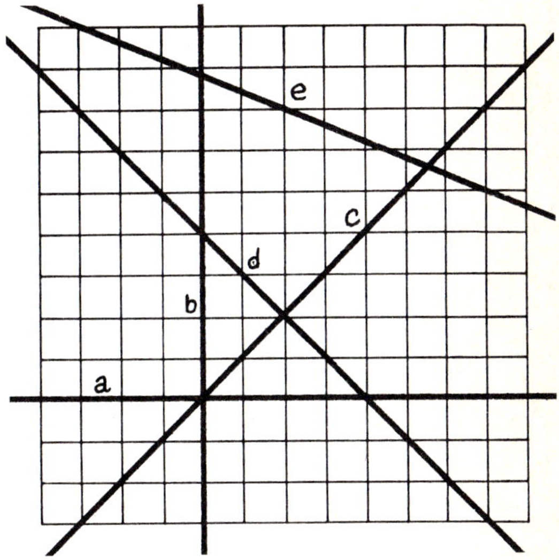

Figure 17

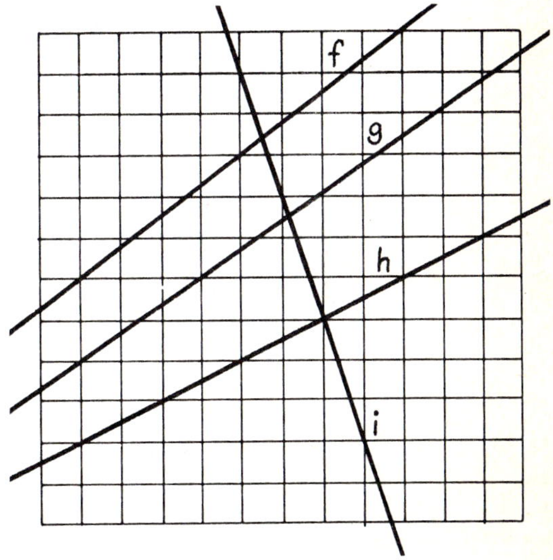

Figure 18

3. From the point P in Figure 19, draw lines a, b, c and d with the given slopes. Identify each line.

Line	Slope
a	1
b	-1
c	2
d	-2

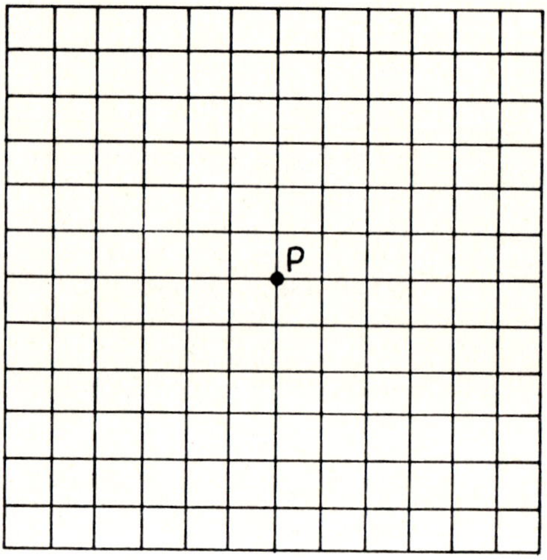

Figure 19

4. From the point P in Figure 20, draw lines e, f, g and h with the given slopes. Identify each line.

Line	Slope
e	$\frac{1}{3}$
f	$-\frac{1}{3}$
g	$\frac{2}{5}$
h	$-\frac{2}{5}$

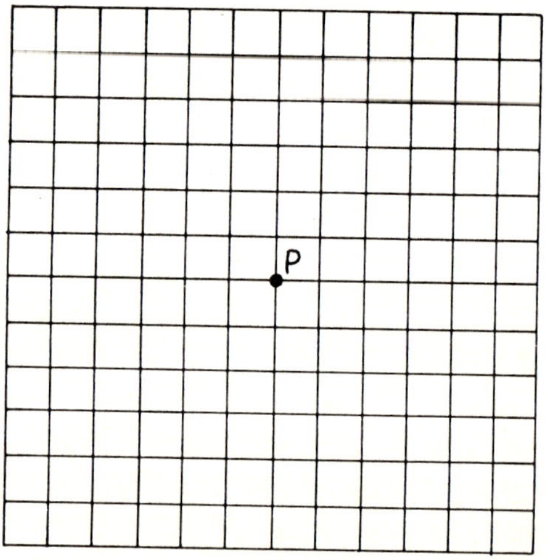

Figure 20

5. From the point P in Figure 21, draw lines i, j, k, l and m with the given slopes. Identify each line.

Line	Slope
i	0
j	$\frac{3}{2}$
k	$-\frac{3}{2}$
l	$\frac{4}{5}$
m	0.5

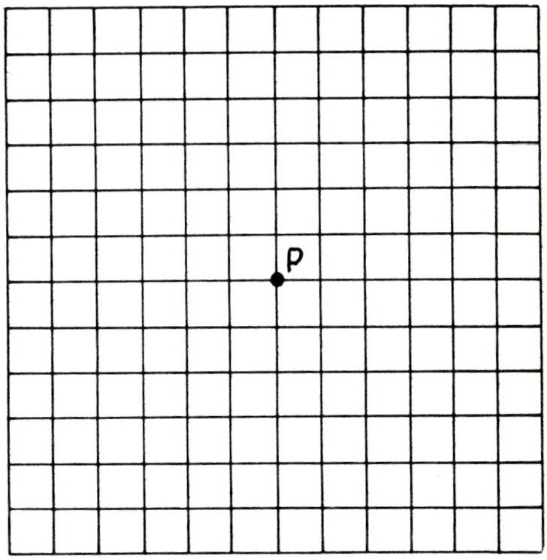

Figure 21

Answers to Exercises

1. a, 3
   b, 2
   c, 2
   d, undefined
   e, 0
   f, -1
   g, $\frac{-3}{2}$
   h, -2

2. a, 0
   b, undefined
   c, 1
   d, -1
   e, $\frac{-2}{5}$
   f, $\frac{3}{4}$
   g, $\frac{2}{3}$
   h, $\frac{1}{2}$
   i, -3

3.

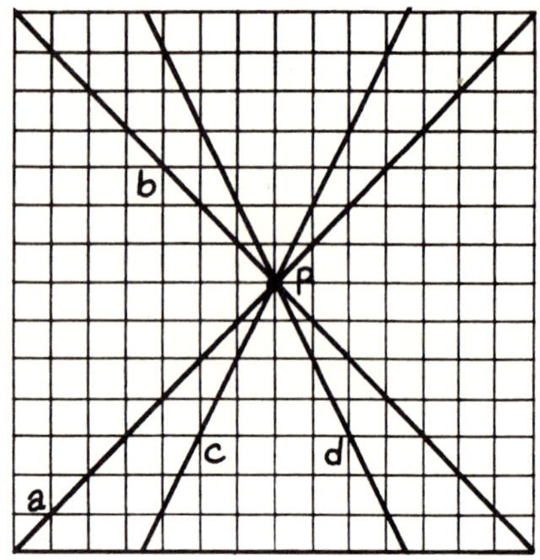

4.

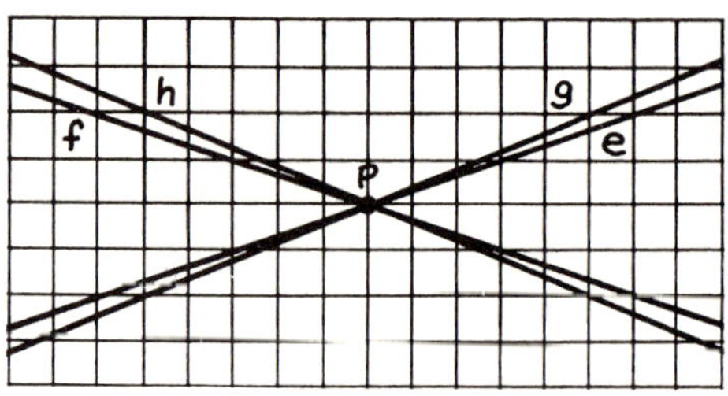

5.

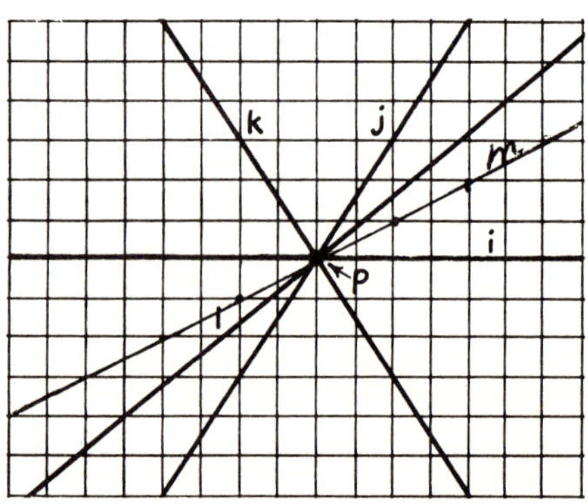

Additional Exercises

1. Calculate the slope of each of the lines in Figures 22 and 23.

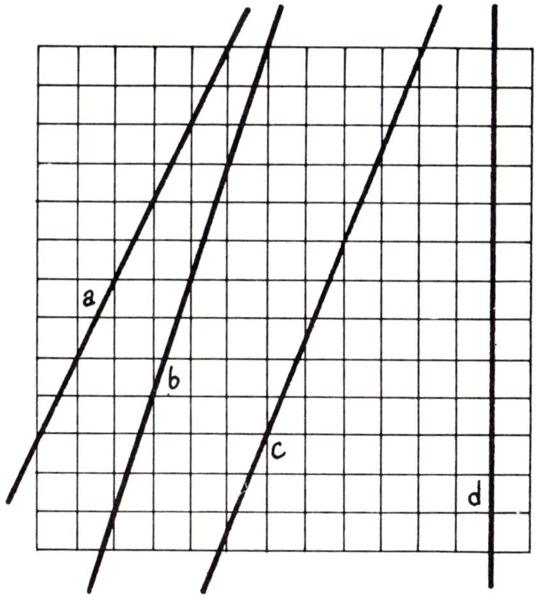

Figure 22

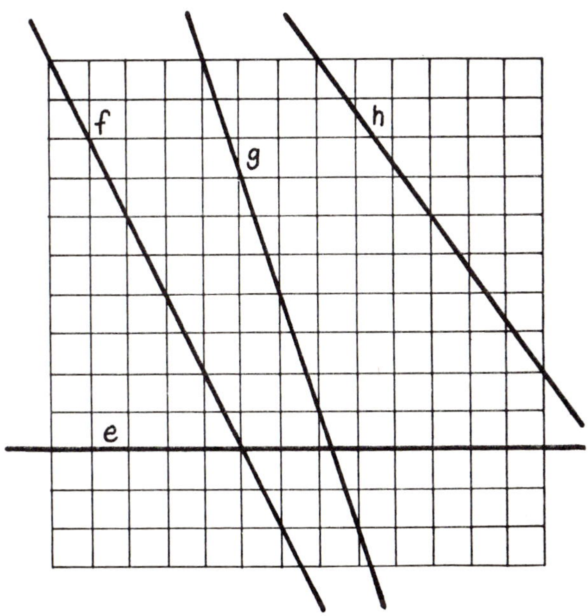

Figure 23

2. Match each line in Figures 24 and 25 with one of the given slopes.

Slope
$\frac{3}{4}$
-2
-1
0
undefined
$\frac{-3}{2}$
$\frac{2}{3}$
$\frac{1}{5}$

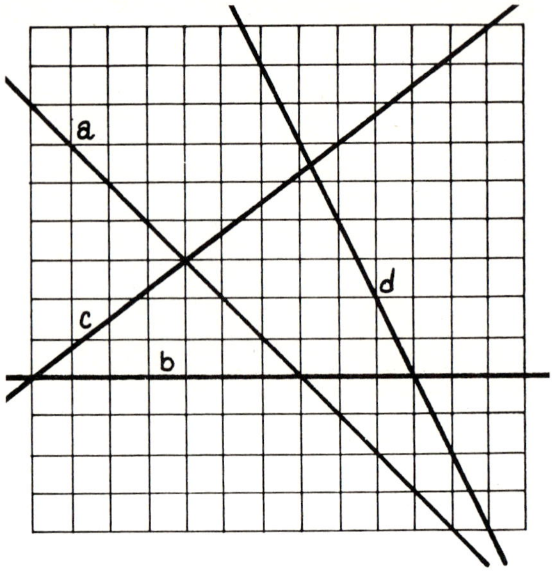

Figure 24

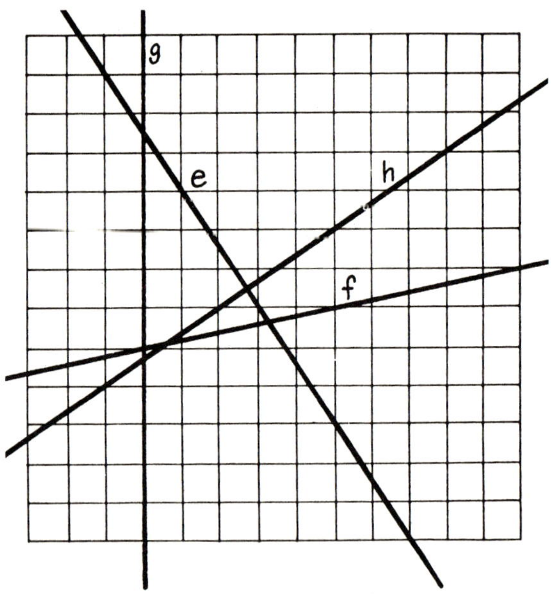

Figure 25

3. From the point P in Figure 26, draw lines a, b, c and d with the given slopes. Identify each line.

Line	Slope
a	2
b	-2
c	$\frac{1}{2}$
d	$-\frac{1}{2}$

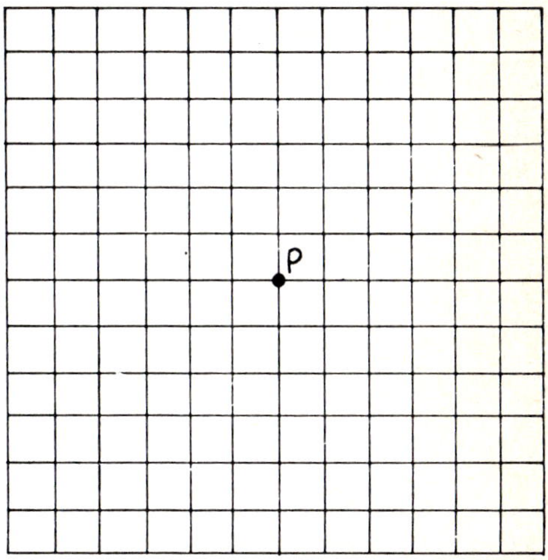

Figure 26

4. From the point P in Figure 27 draw lines e, f, g and h with the given slopes. Identify each line.

Line	Slope
e	$\frac{3}{5}$
f	$-\frac{3}{5}$
g	0
h	undefined

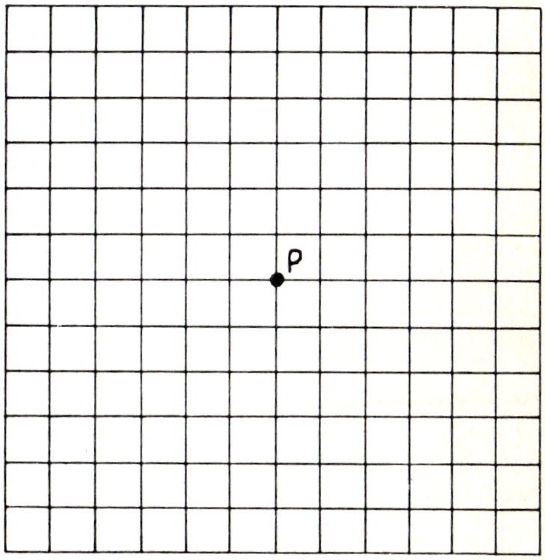

Figure 27

5. From the point P in Figure 28 draw lines i, j, k, l and m with the given slopes. Identify each line.

Line	Slope
i	$-\frac{5}{2}$
j	$\frac{5}{2}$
k	$\frac{2}{3}$
l	$\frac{-3}{2}$
m	0.4

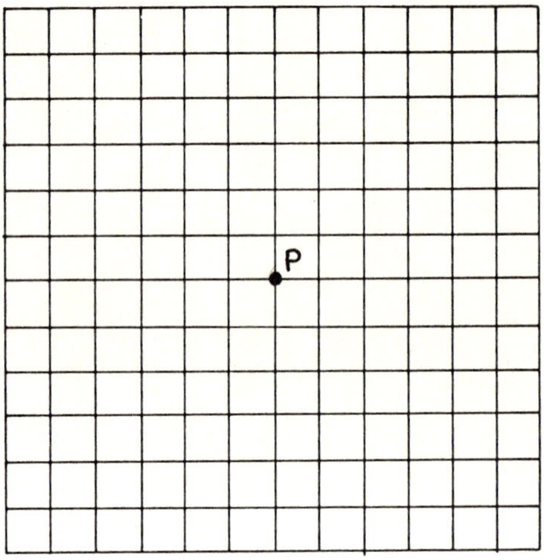

Figure 28

Lesson 6

The Intercept–Slope Form

In Lesson 4, we graphed the solutions of an equation by finding two solutions, plotting them as points and drawing a line through them. In this section we are going to graph the solutions of an equation using slopes.

EXAMPLE 1 Graph the solutions of $y = 2x + 1$.

If we plot $y = 2x + 1$ the way we did in lesson 4 we get the graph in Figure 1.

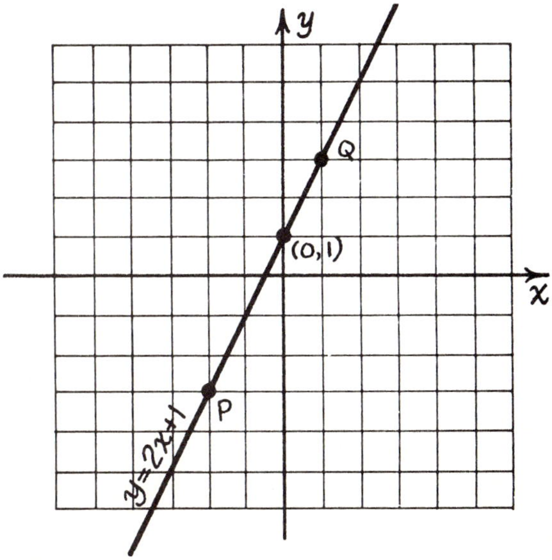

Figure 1

Note

(1) The line crosses the y-axis at (0, 1). This point is called the **y-intercept.**

(2) The slope of the line is 2, since if we pick any two points on the line, say P and Q, we find:

$$\text{Slope} = \frac{\Delta y}{\Delta x} = \frac{6}{3} = 2$$

(3) In the equation of the line $y = 2x + 1$, there is a 1 at the end of the equation. Remember the y-intercept is (0, 1). Also there is a 2 multiplying the x. Remember, we found that the slope of the line is 2.

EXAMPLE 2 Graph the solutions of $y = -3x + 2$.

If we plot $y = -3x + 2$ we get the graph in Figure 2.

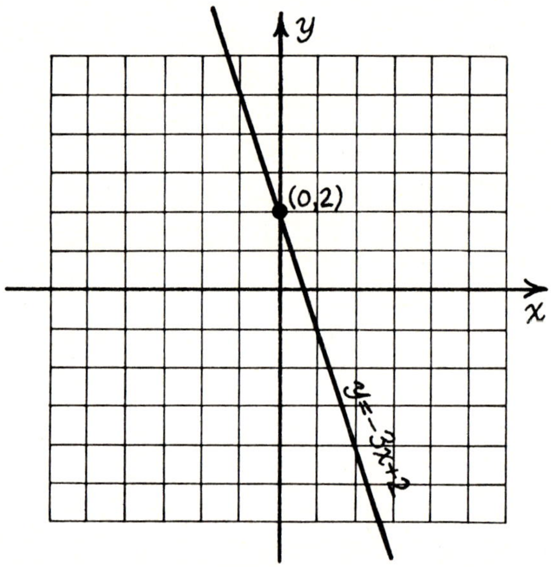

Figure 2

Note

(1) The line crosses the y-axis at (0, 2), so the y-intercept is (0, 2).

(2) The slope of the line is -3. (This can be found by picking any two points on the line and determining Δy and Δx.)

(3) In the equation $y = -3x + 2$, there is a 2 at the end of the equation. Remember the y-intercept is (0, 2). Also there is a -3 multiplying the x. Remember, we found that the slope of the line is -3.

EXAMPLE 3 Graph $y = \frac{1}{2}x - 3$.

If we plot $y = \frac{1}{2}x - 3$, we get the graph in Figure 3.

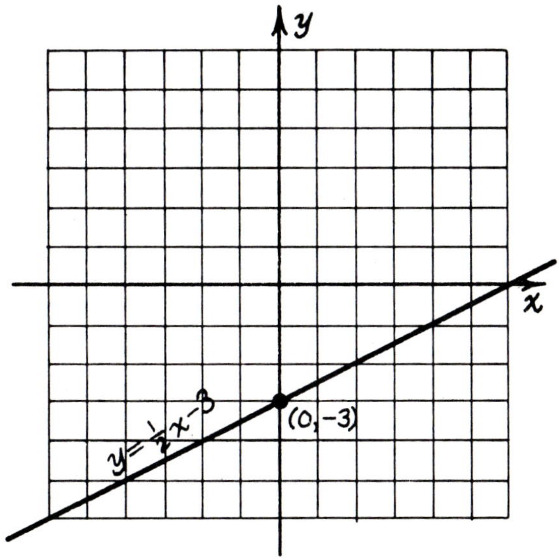

Figure 3

Note

(1) The <u>y-intercept</u> is $(0, -3)$.

(2) The <u>slope</u> of the line is $\frac{1}{2}$.

(3) In the equation $y = \frac{1}{2}x - 3$, there is a -3 at the end of the equation and the y-intercept is $(0, -3)$. Also, there is a $\frac{1}{2}$ multiplying the x and the slope is $\frac{1}{2}$.

EXAMPLE 4 Graph the solutions of $y = 2x$.

If we plot $y = 2x$, we get the graph in Figure 4.

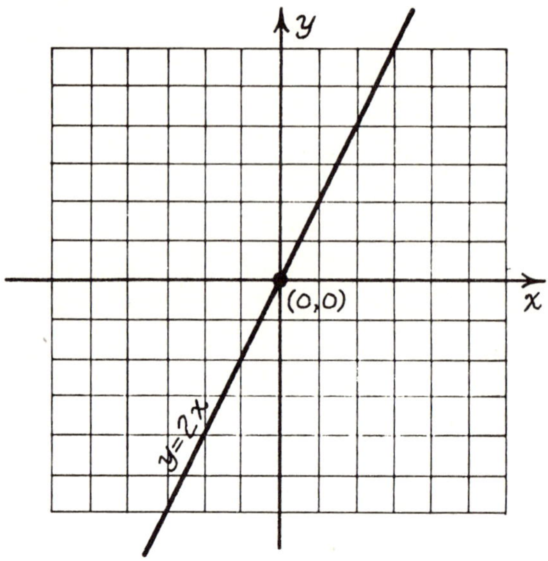

Figure 4

Note

(1) The y-intercept is (0, 0).

(2) The slope is 2.

(3) In the equation $y = 2x$, we can think of a zero at the end of the equation (think: $y = 2x + 0$) and the y-intercept is (0, 0). Also, there is a 2 multiplying the x, and the slope is 2.

EXAMPLE 5 Graph the solutions of $y = x - \frac{1}{2}$.

If we plot $y = x - \frac{1}{2}$, we get the graph in Figure 5.

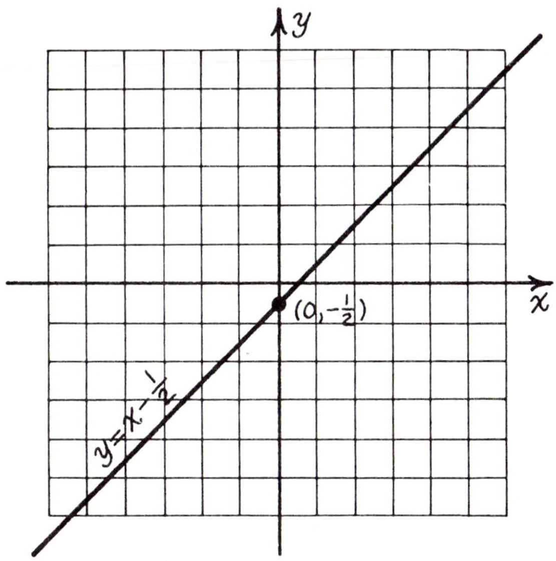

Figure 5

Note

(1) The y-intercept is $\left(0, -\frac{1}{2}\right)$.

(2) The slope is 1.

(3) In the equation $y = x - \frac{1}{2}$, there is a $-\frac{1}{2}$ at the end of the equation and the y-intercept is $\left(0, -\frac{1}{2}\right)$. We can think of a 1 multiplying the x. Think: $y = 1x - \frac{1}{2}$ and the slope is 1.

Let's examine the patterns of Examples 1 through 5. We will use b to stand for the number at the end of the equation and m to stand for the number multiplying the x.

Table 1

Example	Equation	b	y-intercept	m	Slope
1	$y = 2x + 1$	1	(0, 1)	2	2
2	$y = -3x + 2$	2	(0, 2)	−3	−3
3	$y = \frac{1}{2}x - 3$	−3	(0, −3)	$\frac{1}{2}$	$\frac{1}{2}$
4	$y = 2x$	0	(0, 0)	2	2
5	$y = x - \frac{1}{2}$	$-\frac{1}{2}$	$\left(0, -\frac{1}{2}\right)$	1	1
Summary	$y = mx + b$	b	$(0, b)$	m	m

The last line of table 1 shows that whenever an equation is in the form of $y = mx + b$, the y-intercept of its graph is always $(0, b)$ and the slope is always the number m.

We now have a way of writing an equation for any straight line: If a line is vertical, it has no slope. Its equation has the form $x = k$, where k stands for one number. Every point on the line has k as its x-coordinate. See Lesson 3, Example 4.

If a line is not vertical, it has a slope. Its equation has the form $y = mx + b$.

Vertical Line has equation $x = k$
Line not vertical has equation $y = mx + b$

EXAMPLE 6 What is the slope and y-intercept of the graph of $y = -0.5x + 7.2$?

Note that this equation is in the form of $y = mx + b$.

$$y = -0.5x + 7.2$$

$$y = mx + b$$

Since m is -0.5, the slope is -0.5. Since b is 7.2, the y-intercept is (0, 7.2).

EXAMPLE 7 What is the slope and y-intercept of the graph of $2x + 3y = -5$?

First we put the question in the form $y = mx + b$.

$$\begin{array}{lrcl} & 2x + 3y & = & -5 \\ \text{Add } -2x & -2x & & -2x \\ \hline & 3y & = & -2x - 5 \end{array}$$

Divide by 3.

$$\frac{\overset{1}{\cancel{3}}y}{\underset{1}{\cancel{3}}} = \frac{-2x - 5}{3}$$

$$y = -\frac{2}{3}x - \frac{5}{3}$$

This equation is in the form $y = mx + b$.

$$y = -\frac{2}{3}x - \frac{5}{3}$$

Since m is $-\frac{2}{3}$, the slope is $-\frac{2}{3}$. Since b is $-\frac{5}{3}$, the y-intercept is $\left(0, -\frac{5}{3}\right)$.

EXAMPLE 8 Graph the solutions of the equation $y = \frac{2}{3}x - 4$ using the y-intercept and the slope.

Note: This equation is in the form $y = mx + b$, where $b = -4$ and $m = \frac{2}{3}$. So the y-intercept is $(0, -4)$ and the slope is $\frac{2}{3}$.

First, plot the y-intercept $(0, -4)$ as in Figure 6.

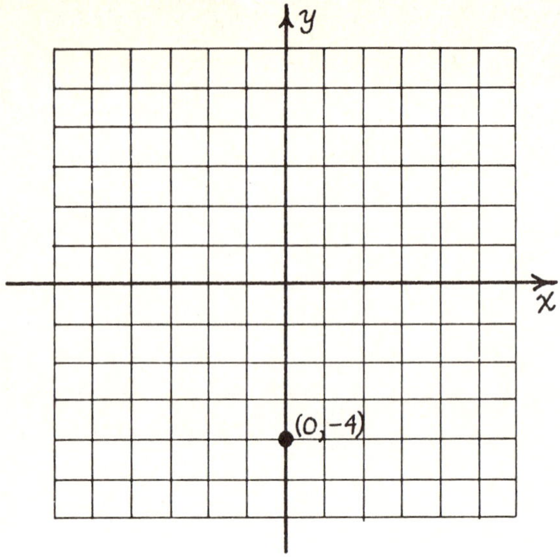

Figure 6

We want the line through this point (0, −4) with slope $\frac{2}{3}$.

$$\text{Slope} = \frac{2}{3} = \frac{\Delta y}{\Delta x}$$

Start at the y-intercept (0, −4), go up 2 units (since $\Delta y = 2$) and go to the right 3 units (since $\Delta x = 3$). Make a dot and draw the line through the two points, as in Figure 7.

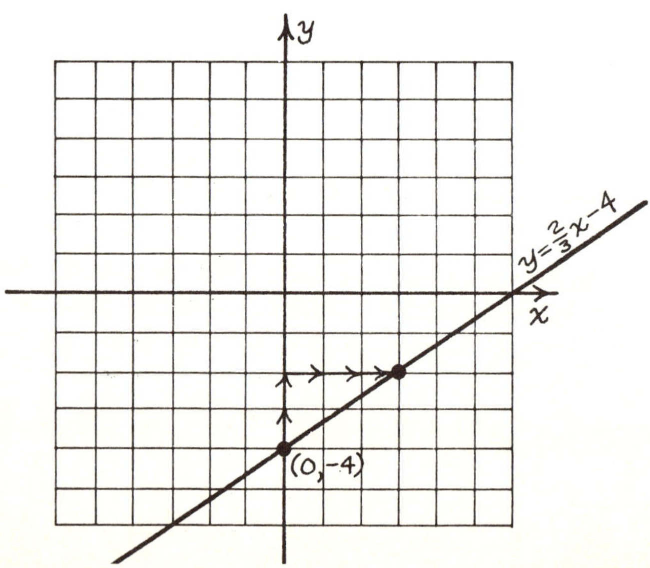

Figure 7

EXAMPLE 9 Graph the solutions of the equation $y = -x + 5$, using the y-intercept and the slope.

Note: This equation is in the form $y = mx + b$, where $b = 5$ and $m = -1$ (remember: $-x = -1x$), so the y-intercept is (0, 5) and the slope is -1.

First, plot the y-intercept (0, 5) as in Figure 8.

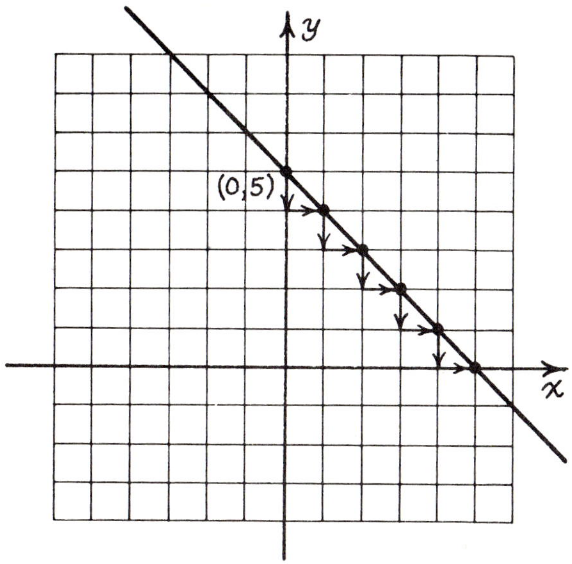

Figure 8

We want the line through this point (0, 5) with slope -1.

$$\text{Slope} = -1 = \frac{-1}{1} = \frac{\Delta y}{\Delta x}$$

Start at the y-intercept (0, 5) go down 1 unit (since $\Delta y = -1$) and go to the right 1 unit (since $\Delta x = 1$). Make a dot. Since the two dots are so close together, repeat the process from the last point drawn to get a few more points. Then draw the line as in Figure 8.

EXAMPLE 10 The graph of an equation is a straight line with slope -7 and y-intercept $(0, 2)$. Write an equation for this line.

Since this equation has a slope (which is -7), the equation can be written in the form $y = mx + b$.

The slope is -7, so m is -7
The y-intercept is $(0, 2)$, so b is 2

So an equation of the line is $y = -7x + 2$.

EXAMPLE 11 The graph of an equation is a straight line with slope zero and y-intercept $(0, -3)$. Write an equation for this line and graph the line.

Since this line has a slope (which is zero), the equation can be written in the form $y = mx + b$.

The slope is 0, so m is 0
The y-intercept is $(0, -3)$, so b is -3

So an equation of the line is $y = 0x - 3$, which is $y = -3$. The line is sketched in Figure 9.

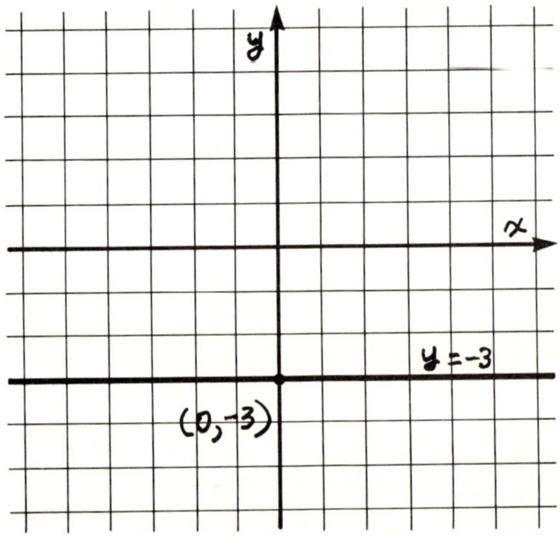

Figure 9

EXAMPLE 12 The graph of an equation is a straight line whose slope is undefined. The line passes through the point (4, 3). Write an equation of this line and graph the line.

Since the slope of this line is undefined, the line has no slope and we have no value for m. This is a vertical line and its equation has the form $x = k$. Since the point (4, 3) is on the line, k is 4. Thus, an equation of this line is $x = 4$. The graph is sketched in Figure 10.

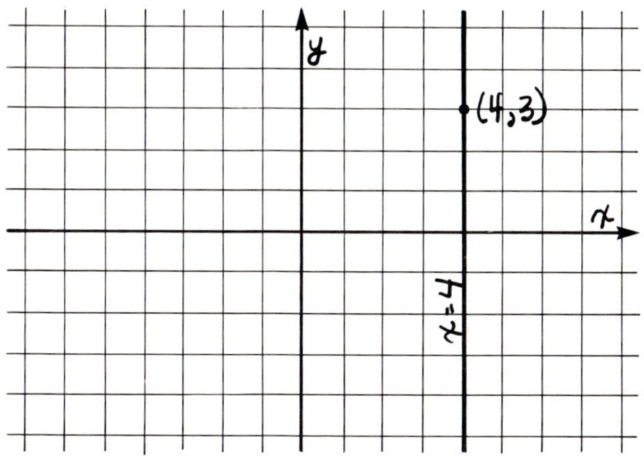

Figure 10

EXAMPLE 13 Write an equation for the straight line graph in Figure 11.

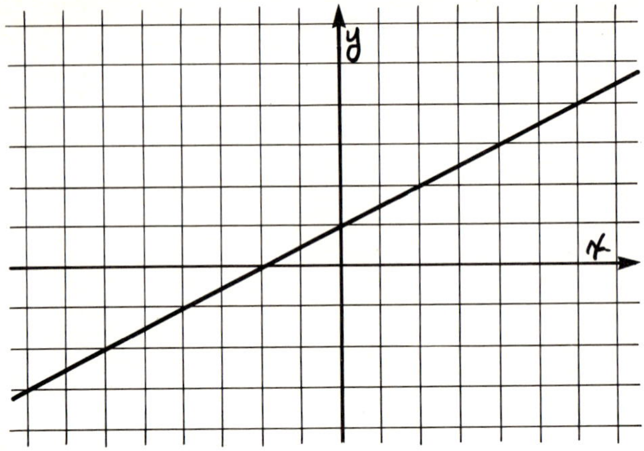

Figure 11

We calculate the slope of this line to be $\frac{1}{2}$, so m is $\frac{1}{2}$. We see that the y-intercept is (0, 1), so b is 1. See Figure 12.

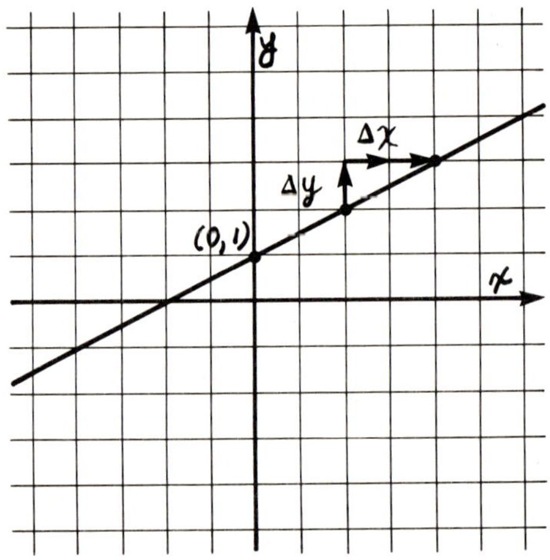

Figure 12

Since we know m and b, the equation

$$y = mx + b$$

becomes $y = \frac{1}{2}x + 1$

Exercises

1. Find the y-intercept and the slope of the graph of each of the following equations.
   a. $y = 3x + 2$
   b. $y = -3x + 5$
   c. $y = \frac{2}{3}x - 2$
   d. $y = -x + 1$
   e. $y = x - 4$
   f. $y = -x$
   g. $y = -\frac{4}{5}x$
   h. $y = 2x - 3$
   i. $y = -2x + 1$
   j. $y = -\frac{5}{3}x + 1$
   k. $y = 3x + \frac{1}{2}$
   l. $y = \frac{2x}{5} - 4$

2. Graph the solutions of each of the equations in example 1, using the y-intercept and the slope. Label each with its equation.

3. What is the slope and y-intercept of the graph of $y = 0.5x + 4.2$?

4. What is the slope and y-intercept of the graph of $y = -1.5x - 5.4$?

5. What is the slope and y-intercept of the graph of $3x + 2y = 6$?

6. What is the slope and y-intercept of the graph of $4x - 3y = 12$?

7. What is the slope and y-intercept of the graph of $x - 4y = 2$?

8. What is the slope and y-intercept of the graph of $2x - y = 8$?

9. Given the slope and the y-intercept , find the equation for each of the following lines. Graph each line.
   a. slope is -2 and y-intercept is $(0, 3)$
   b. slope is 1 and y-intercept is $(0, -1)$
   c. slope is zero and y-intercept is $(0, -2)$
   d. slope is $-\frac{1}{2}$ and y-intercept is $(0, 1)$
   e. slope is $\frac{2}{3}$ and y-intercept is $(0, -4)$

10. Write an equation for each of the straight line graphs given below.

a.

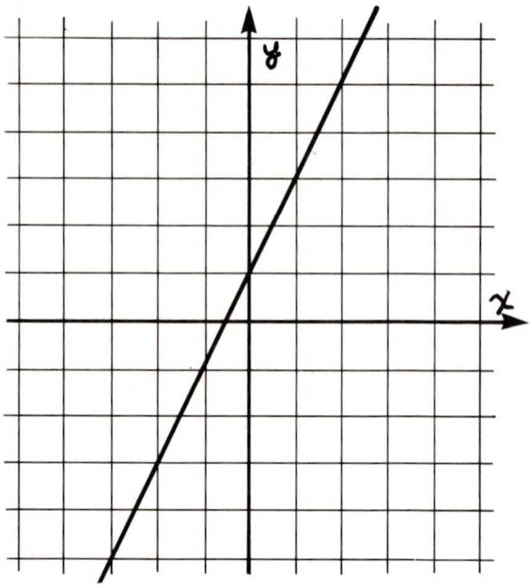

b.

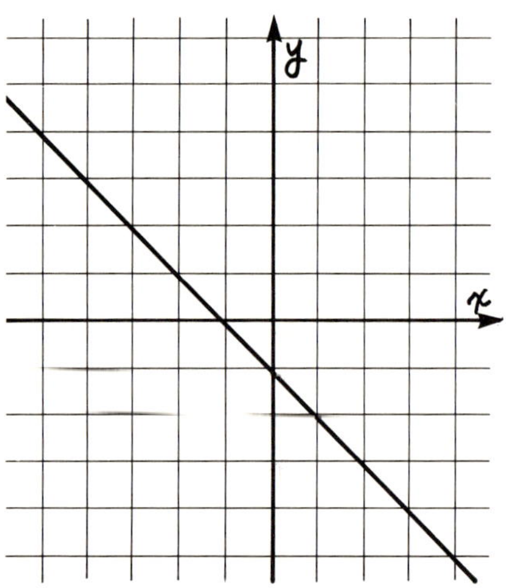

c.

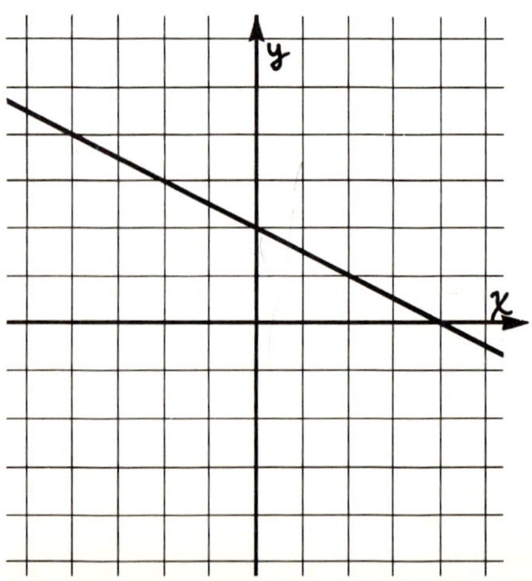

Answers to Exercises

1. a. (0,2) $m = 3$
   b. (0,5) $m = -3$
   c. (0,-2) $m = \frac{2}{3}$
   d. (0,1) $m = -1$
   e. (0,-4) $m = 1$
   f. (0,0) $m = -1$
   g. (0,0) $m = -\frac{4}{5}$
   h. (0,-3) $m = 2$
   i. (0,1) $m = -2$
   j. (0,1) $m = \frac{-5}{3}$
   k. $(0,\frac{1}{2})$ $m = 3$
   l. (0,-4) $m = \frac{2}{5}$

2a) $y = 3x + 2$

y-intercept (0,2)

slope = 3

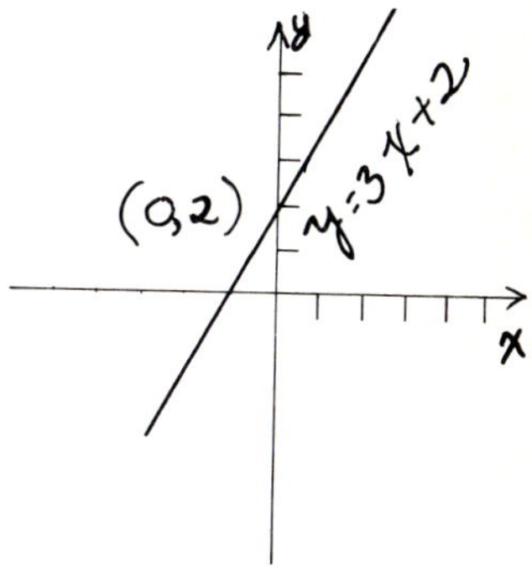

2b) $y = -3x + 5$

y-intercept (0,5)

slope = -3

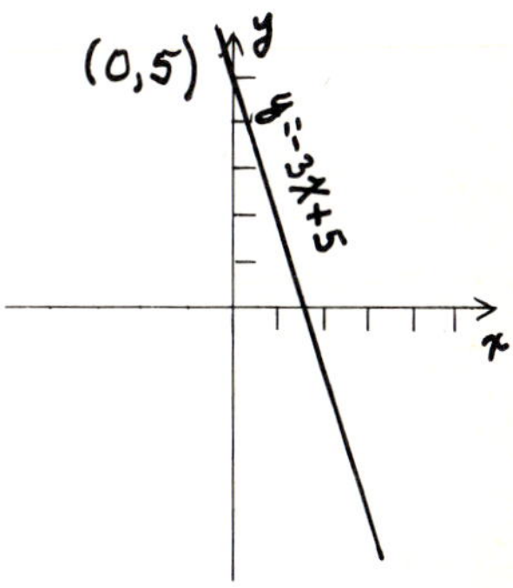

2c) $y = \frac{2}{3}x - 2$

(0, −2)

$m = \frac{2}{3}$

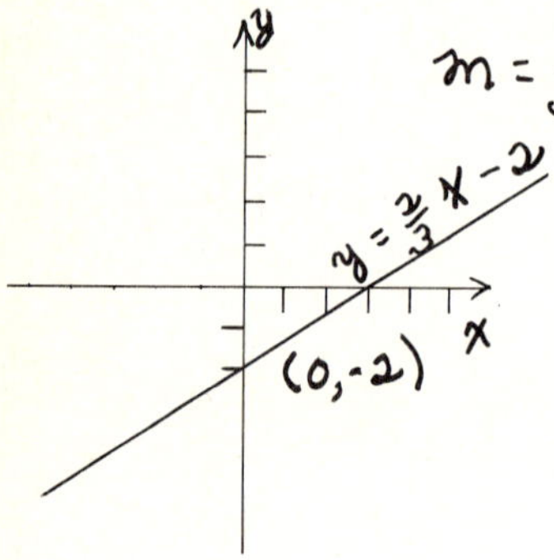

2d) $y = -x + 1$

(0, 1)

$m = -1$

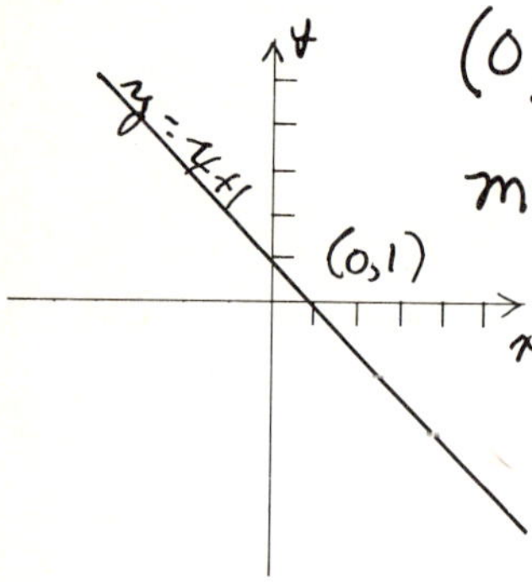

2e) $y = x - 4$

(0, −4)

$m = 1$

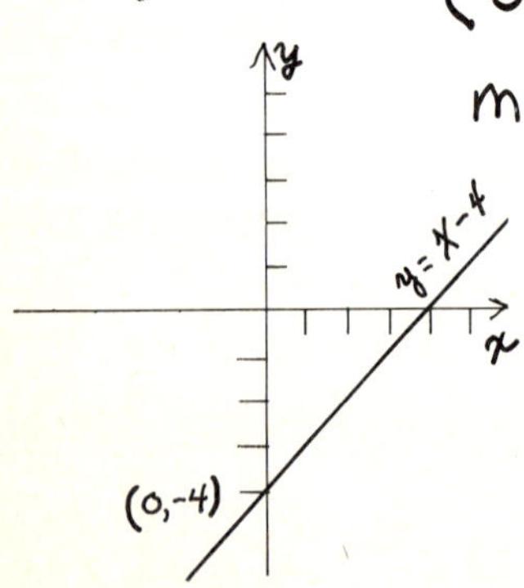

2f) $y = -x$

(0, 0)

$m = -1$

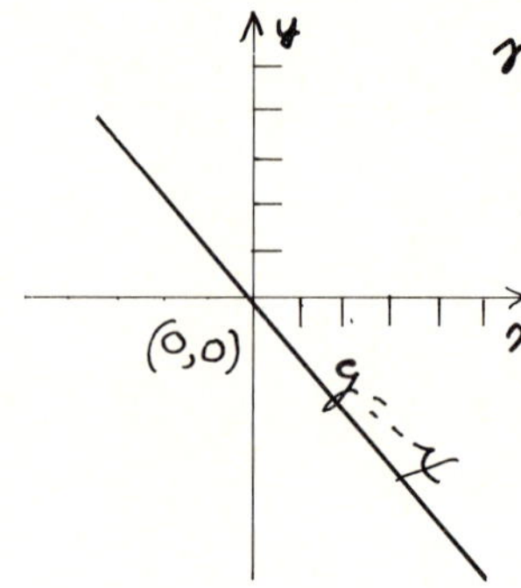

2g) $y = -\frac{4}{5}x$

(0, 0)

$m = -\frac{4}{5}$

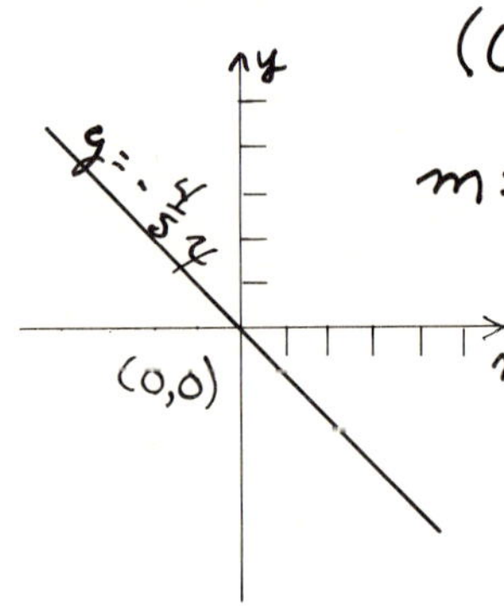

2h) $y = 2x - 3$

(0, −3)

$m = 2$

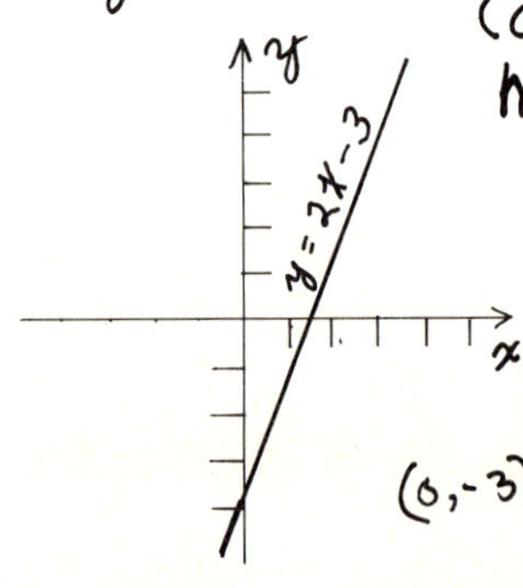

2i) $y = -2x + 1$

(0,1)

$m = -2$

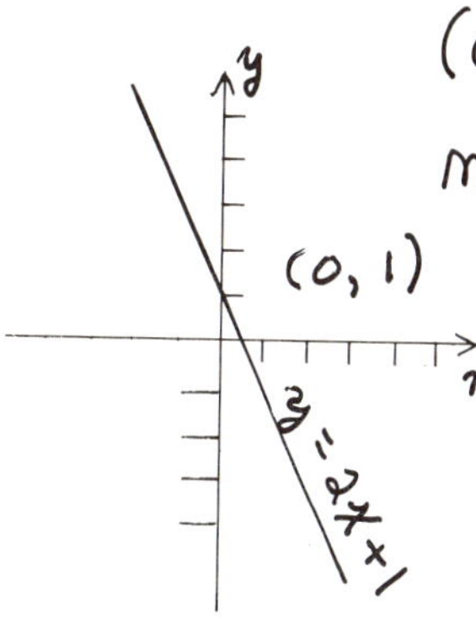

2l) $y = \frac{2}{5}x - 4$

(0, −4)

$m = \frac{2}{5}$

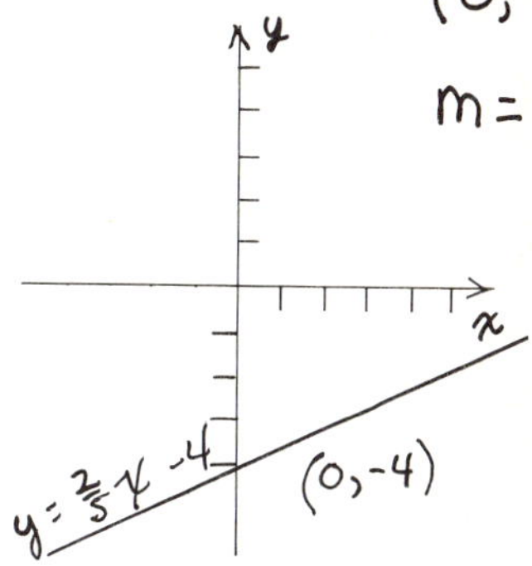

2j) $y = -\frac{5}{3}x + 1$

(0, 1)

$m = -\frac{5}{3}$

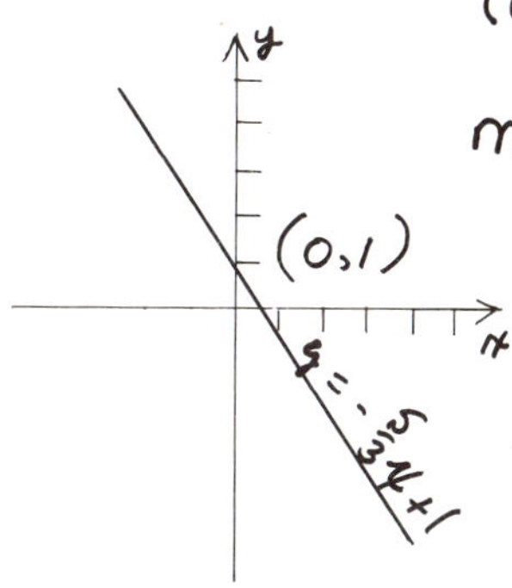

2k) $y = 3x + \frac{1}{2}$

$(0, \frac{1}{2})$

$m = 3$

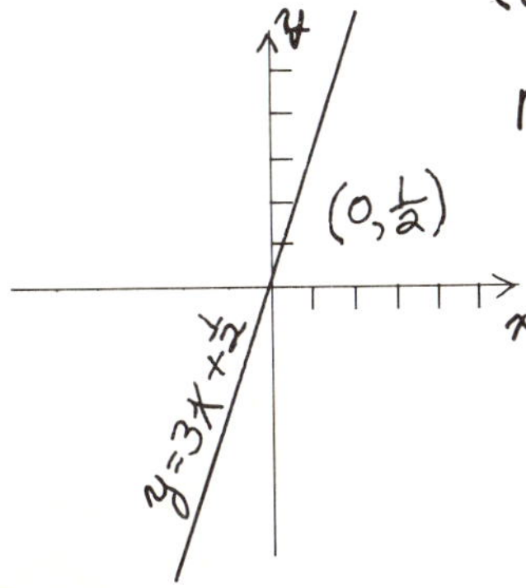

3. $m = 0.5$ (0, 4.2)

4. $m = -1.5$ (0, −5.4)

5. $m = \frac{-3}{2}$ (0, 3)

6. $m = \frac{4}{3}$ (0, −4)

7. $m = \frac{1}{4}$ $(0, -\frac{1}{2})$

8. $m = 2$ (0, −8)

9a

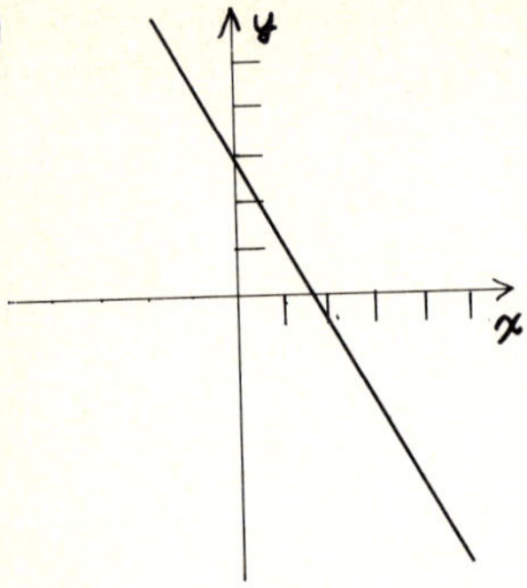

9d

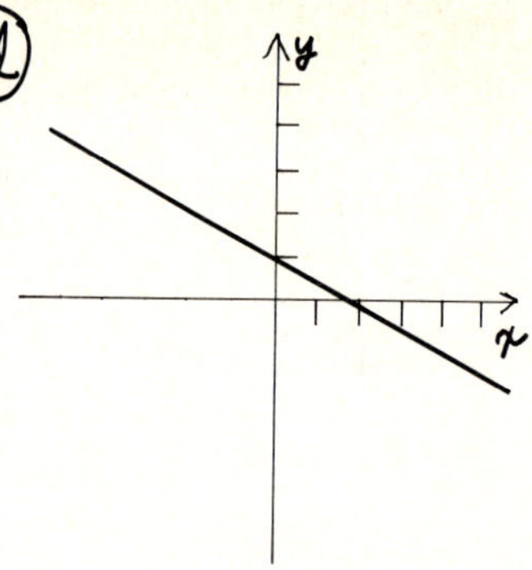

9b

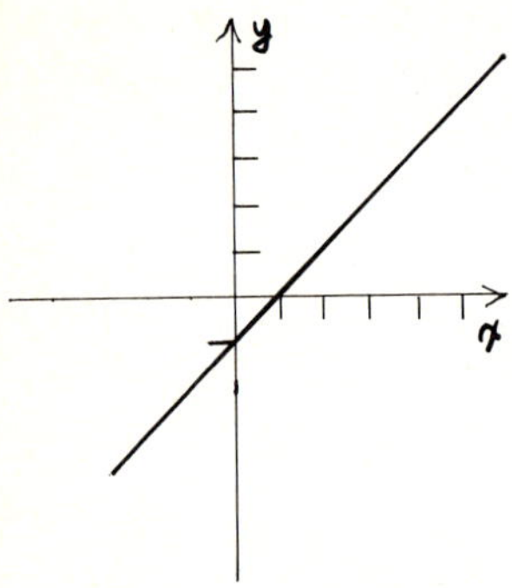

9e

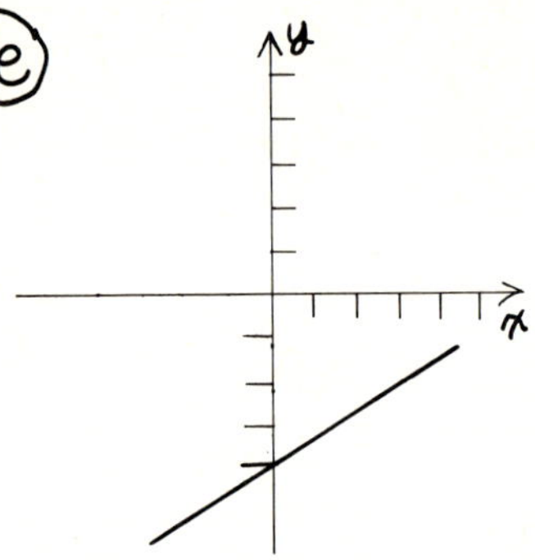

9c

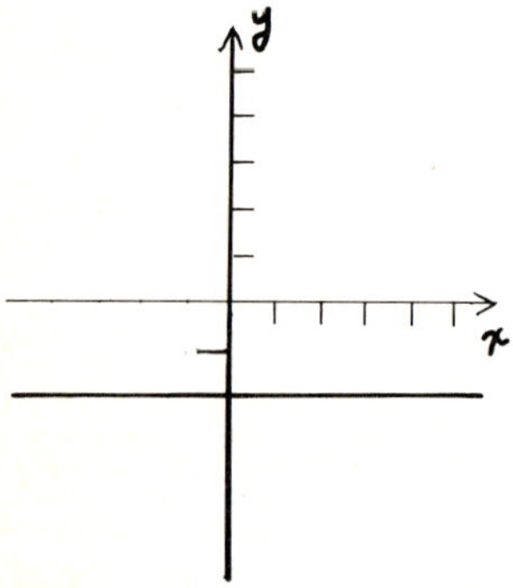

10a $y = 2x + 1$

10b $y = -x - 1$

10c $y = -\frac{1}{2}x + 2$

Additional Exercises

1. Find the y-intercept and the slope of the graph of each of the following equations.
   a. $y = 4x + 1$
   b. $y = -2x + 3$
   c. $y = \frac{3x}{5} - 4$
   d. $y = x - 1$
   e. $y = -x + 1$
   f. $y = -x$
   g. $y = \frac{-3}{4}x$
   h. $y = 2x - 5$
   i. $y = \frac{-4}{3}x + \frac{1}{3}$
   j. $y = 3x - 7$
   k. $y = \frac{-7}{2}x + \frac{7}{2}$
   l. $y = \frac{-3}{5}x + 3$
2. Graph the solutions of each of the equations in Example 1, using the y-intercept and the slope. Label each line with its equation.

3. What is the slope and y-intercept of the graph of $y = 1.5x - 2.7$?

4. What is the slope and y-intercept of the graph of $y = -2.5x - 4.5$?

5. What is the slope and y-intercept of the graph of $4x + 3y = 6$?

6. What is the slope and y-intercept of the graph of $5x - 2y = 10$?

7. What is the slope and y-intercept of the graph of $x - 3y = -1$?

8. What is the slope and y-intercept of the graph of $3x - y = 9$?

9. Given the slope and the y-intercept , find the equation for each of the following lines. Graph each line.
   a. slope is -3 and y-intercept is $(0, 1)$
   b. slope is 2 and y-intercept is $(0, -3)$
   c. slope is zero and y-intercept is $(0, 4)$
   d. slope is $-\frac{1}{3}$ and y-intercept is $(0, 2)$
   e. slope is $\frac{1}{2}$ and y-intercept is $(0, -2)$

10. Write an equation for each of the straight line graphs given below.

a.

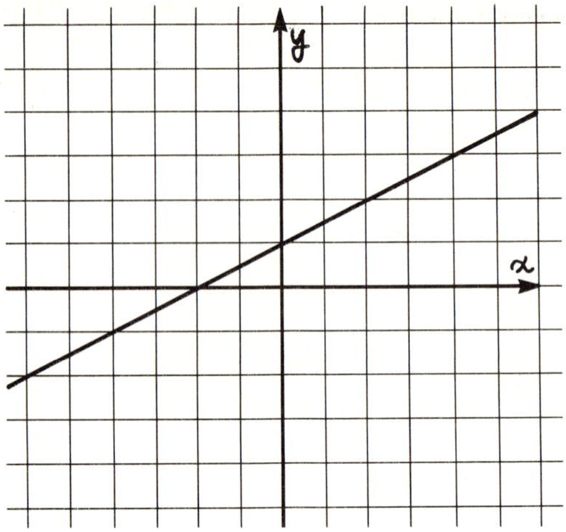

b.

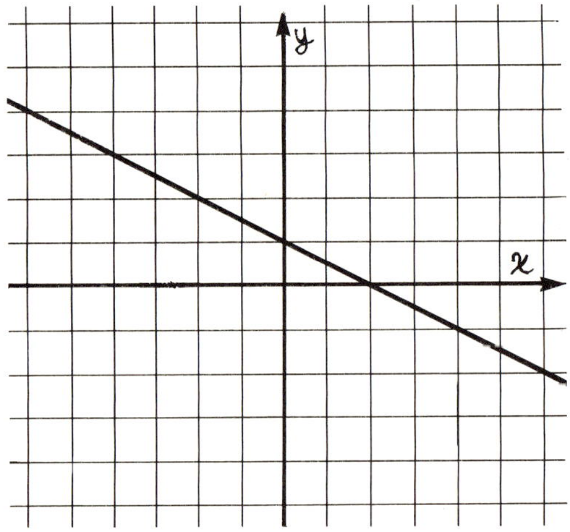

c.

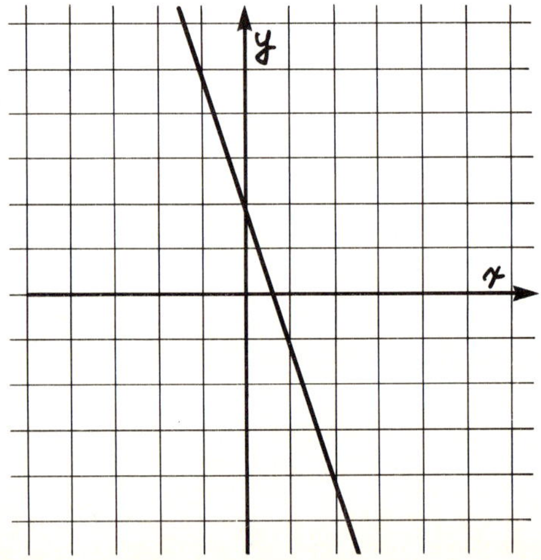

Lesson 7

Simultaneous Solutions of Linear Equations by Graphing

Let's graph the solutions of two different equations on the same picture.

EXAMPLE 1 Graph $y = x + 1$ and $y = -x + 3$ on the same picture.

These lines are graphed in Figure 1.

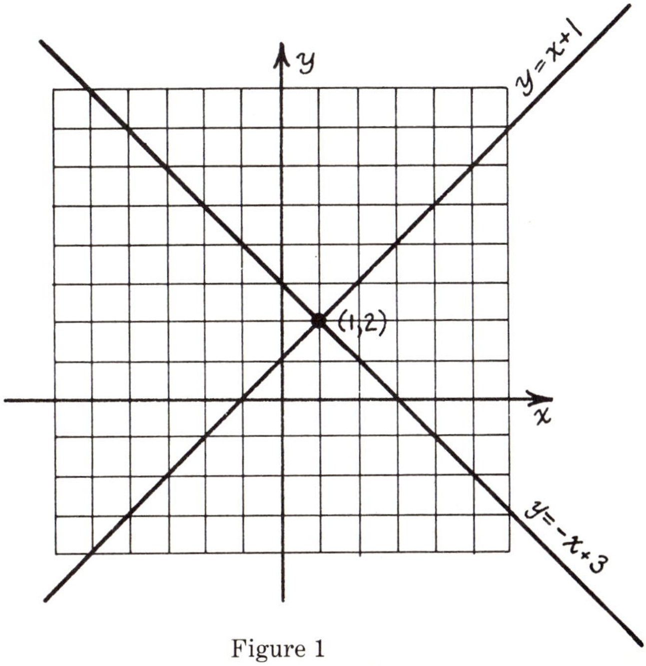

Figure 1

Notice that the two lines cross at the point (1, 2), so (1, 2) is on both lines. Since (1, 2) is on the graph of $y = x + 1$, it is a solution of the equation $y = x + 1$. Also, since (1, 2) is on the graph of $y = -x + 3$, it is a solution of the equation $y = -x + 3$.

We call a point which is the solution of two different equations the **simultaneous solution** of the two equations. Although there are many solutions of the first equation and many solutions of the second equation, (1, 2) is a point that is a simultaneous solution of the two equations. That is, (1, 2) makes $y = x + 1$ true and (1, 2) also makes $y = -x + 3$ true. Since these two lines cross only once, (1, 2) is the only simultaneous solution of the equations $y = x + 1$ and $y = -x + 3$.

EXAMPLE 2 Find the simultaneous solution of the equations $y = 3x$ and $y = -2x - 5$ by graphing.

We have graphed the lines in Figure 2.

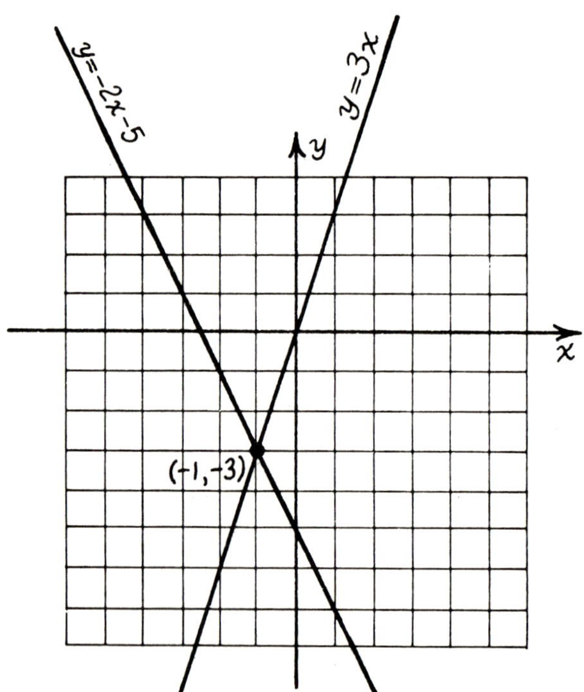

Figure 2

Notice that the two lines cross at the point $(-1, -3)$. So $(-1, -3)$ is a solution of both of the equations; therefore, it is a simultaneous solution of the two equations. These lines cross only once; therefore, $(-1, -3)$ is <u>the</u> simultaneous solution of these equations. To check that $(-1, -3)$ is a solution of both equations we plug in -1 for x and -3 for y as follows.

$$\begin{aligned} y &= 3x \\ -3 &= 3(-1) \\ -3 &= -3 \quad \underline{\text{True}} \end{aligned} \qquad\qquad \begin{aligned} y &= -2x - 5 \\ -3 &= -2(-1) - 5 \\ -3 &= +2 - 5 \\ -3 &= -3 \quad \underline{\text{True}} \end{aligned}$$

So $(-1, -3)$ is the simultaneous solution of both equations.

EXAMPLE 3 Find the simultaneous solution of the equations $y = 2x - 2$ and $y = 0.5x + 1$.

These lines are graphed in Figure 3.

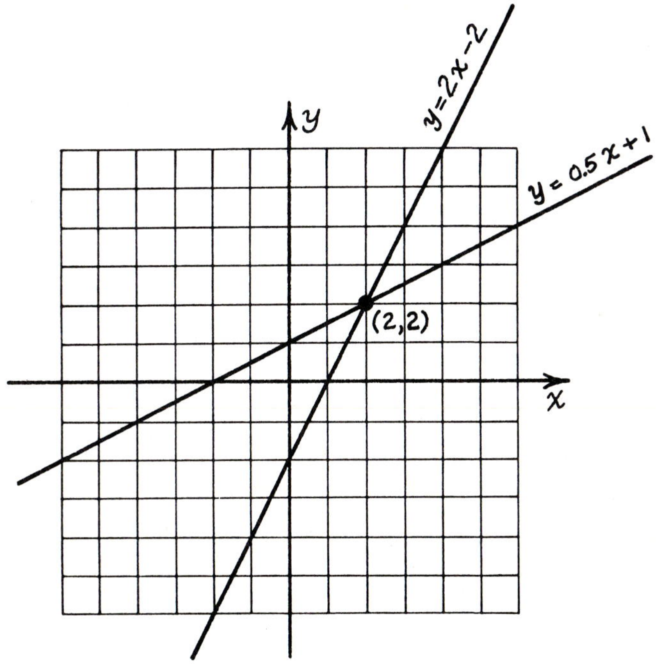

Figure 3

In this case, (2, 2) is the point where the two lines cross. Therefore, (2, 2) is the simultaneous solution of both equations. This means that (2, 2) makes both equations true. Let's check. We plug in 2 for x and 2 for y as follows.

$$\begin{aligned} y &= 2x - 2 \\ 2 &= 2(2) - 2 \\ 2 &= 4 - 2 \\ 2 &= 2 \qquad \underline{\text{True}} \end{aligned} \qquad\qquad \begin{aligned} y &= 0.5x + 1 \\ 2 &= 0.5(2) + 1 \\ 2 &= 1 + 1 \\ 2 &= 2 \qquad \underline{\text{True}} \end{aligned}$$

So (2, 2) is the simultaneous solution of both equations.

EXAMPLE 4 Find the simultaneous solution of the equations $y = 2x + 3$ and $y = 2x - 1$.

These lines are graphed in Figure 4.

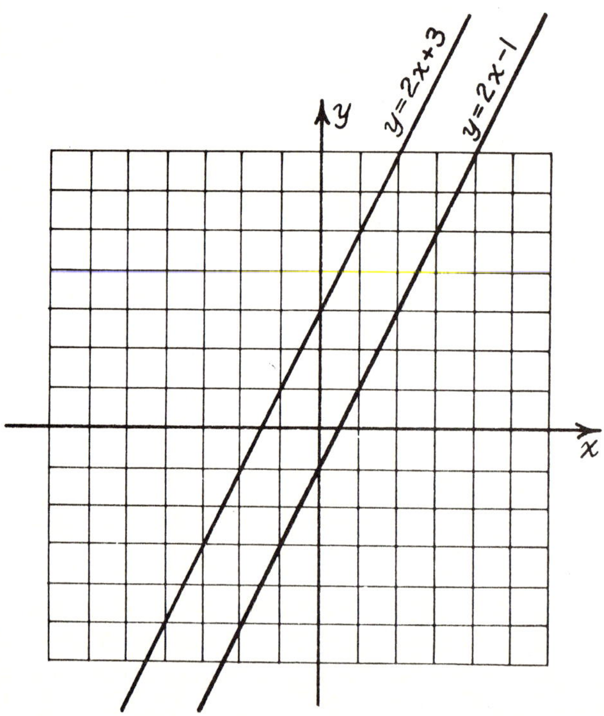

Figure 4

These lines do not cross! They never cross! So even though there are many solutions to the first equation and many solutions to the second equation, there is no simultaneous solution of these two equations. Notice that the slope of line $y = 2x + 3$ is 2 and that the slope of the line $y = 2x - 1$ is also 2. They both have the same slope. When lines have the same slope, we call them **parallel lines.**

EXAMPLE 5 Find the simultaneous solution of the equations $y = -\frac{2}{3}x + 1$ and $y = -\frac{2}{3}x + 2$.

We do not have to draw the graph, since the slope of both lines is $-\frac{2}{3}$ and since the y-intercepts are different, the lines are parallel. Therefore, they do not cross and there is no simultaneous solution of these equations.

EXAMPLE 6 Find the simultaneous solution of $x + y = 7$ and $x - y = 3$.

To graph these we must first rewrite them in intercept–slope form.

$$x + y = 7 \text{ can be written as } y = -x + 7$$
$$x - y = 3 \text{ can be written as } y = x - 3$$

We graph these in Figure 5.

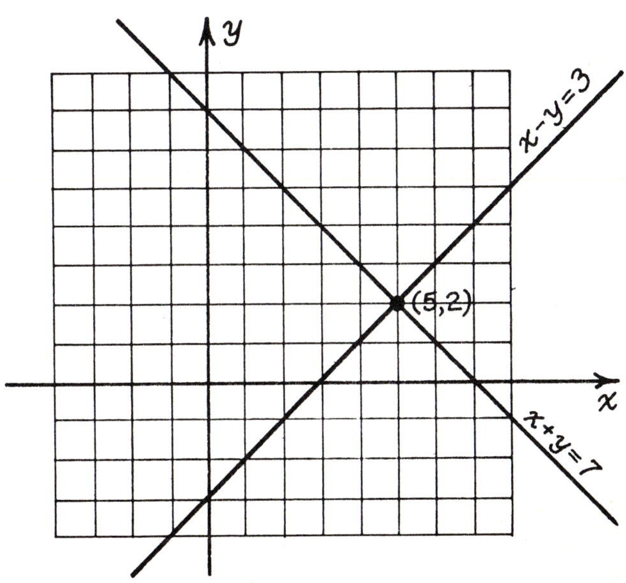

Figure 5

The lines cross at (5, 2), so (5, 2) is the simultaneous solution of these equations. Now let's check by plugging our solution into the two original equations.

$x + y = 7$		$x - y = 3$	
$5 + 2 = 7$		$5 - 2 = 3$	
$7 = 7$	True	$3 = 3$	True

So (5, 2) is the simultaneous solution of both equations.

EXAMPLE 7 Find the simultaneous solution of the equations $y = -\frac{1}{3}x + 4$ and $y = 2x - 2$.

The lines are graphed in Figure 6. We see that the lines cross at a point, but we cannot determine the exact coordinates. The x-coordinate is between 2 and 3, and the y-coordinate is between 3 and 4. In the next lesson we show how to get the exact simultaneous solution of these equations.

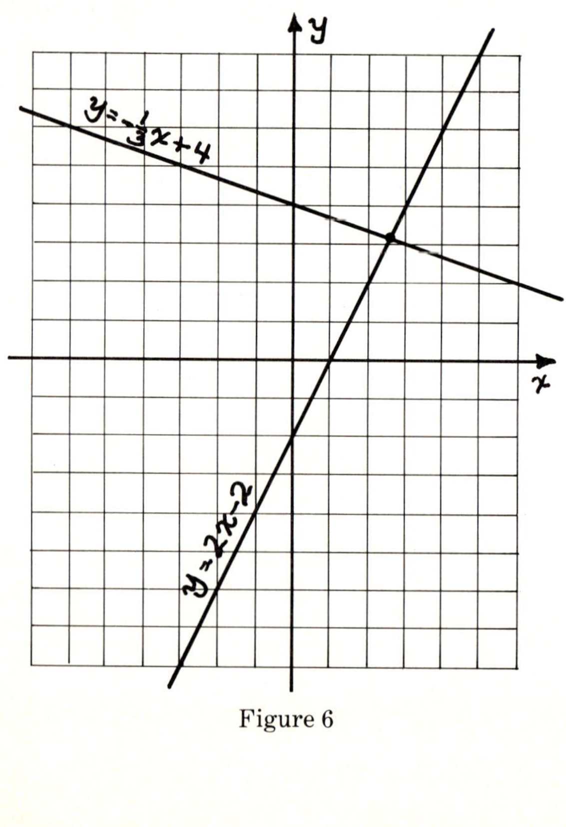

Figure 6

Exercises

Determine the simultaneous solution (if it exists) of each of the following pairs of equations by graphing.

1. $y = -x + 3$
   $y = x - 1$
2. $y = 2x + 3$
   $y = x + 1$
3. $y = \frac{1}{2}x - 2$
   $y = -2x + 3$
4. $y = 2x + 1$
   $y = -x + 4$
5. $y = 3x + 4$
   $y = -x$
6. $y = 5x + 2$
   $y = 5x - 3$
7. $y = \frac{2}{3}x - 1$
   $y = \frac{-1}{3}x - 4$
8. $2x + 3y = 6$
   $x + y = 3$
9. $x + y = 6$
   $x - y = 4$
10. $y - 4x + 4 = 0$
    $y - x - 1 = 0$

Answers to Exercises

(1.) $y = -x + 3$

$y = x - 1$

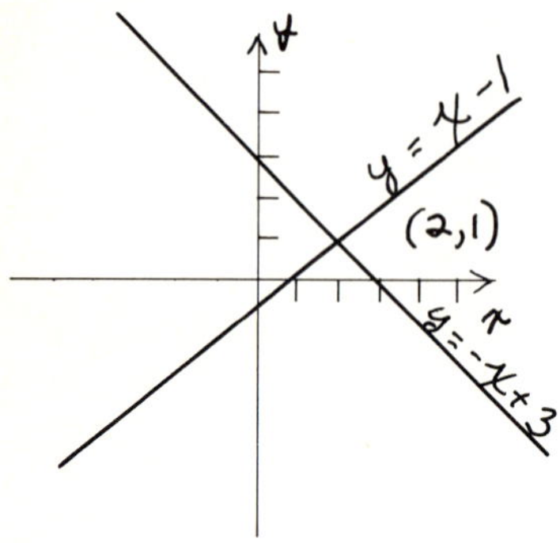

(2.) $y = 2x + 3$

$y = x + 1$

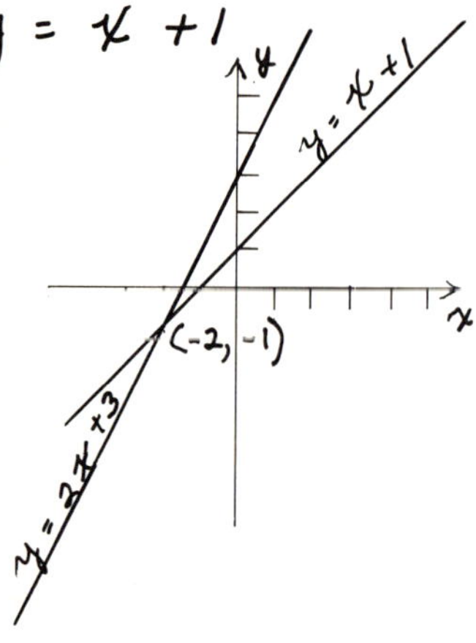

(3.) $y = \frac{1}{2}x - 2$

$y = -2x + 3$

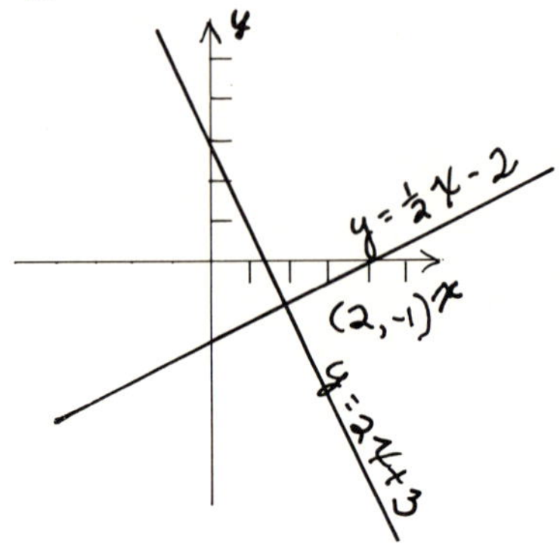

(4.) $y = 2x + 1$

$y = -x + 4$

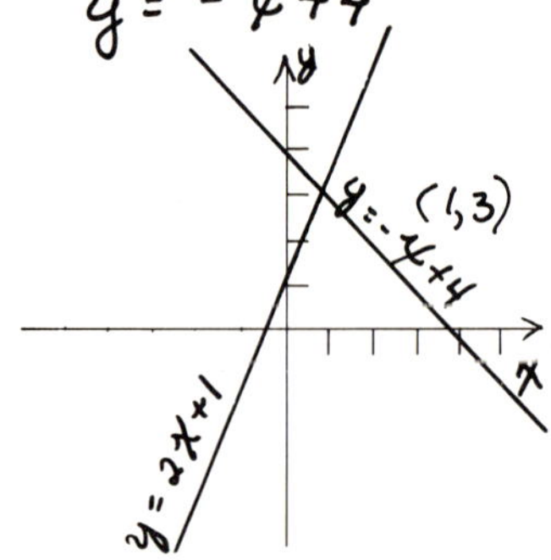

(5.) $y = 3x + 4$

$y = -x$

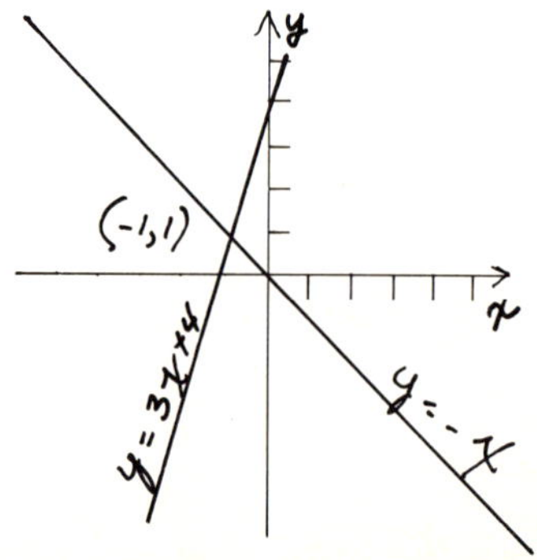

(6.) $y = 5x + 2$

$y = 5x - 3$

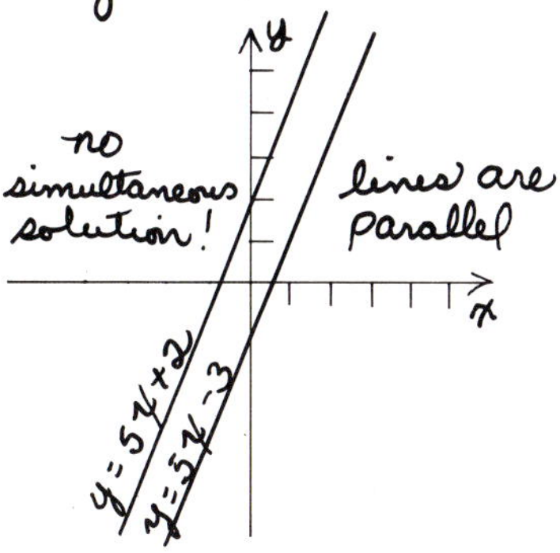

(7.) $y = \frac{2}{3}x - 1$

$y = \frac{-1}{3}x - 4$

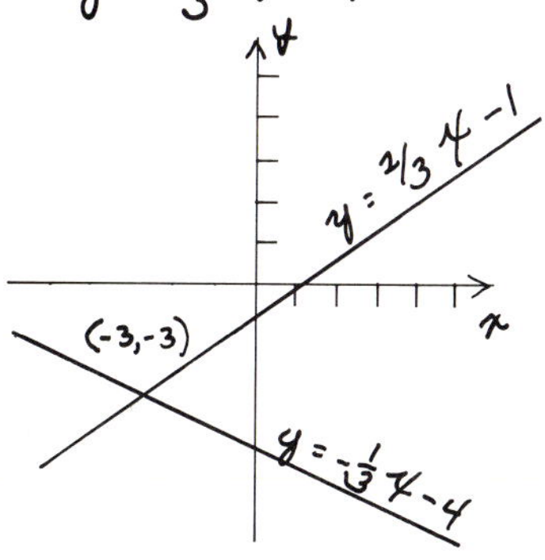

(8.) $3y = 6 - 2x$

$y = 3 - x$

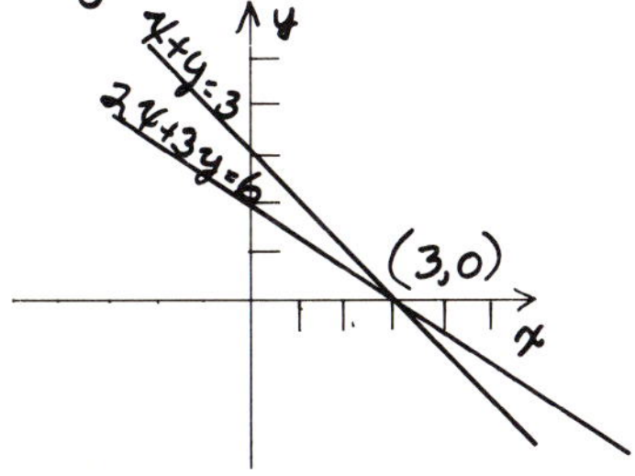

(9.) $y = -x + 6$

$y = x - 4$

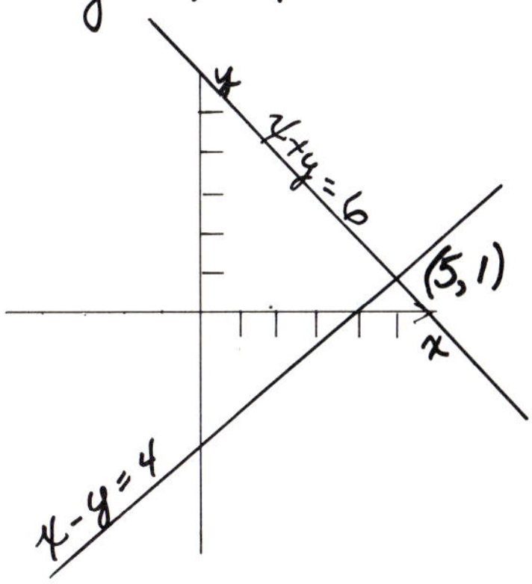

(10.) $y = 4x - 4$

$y = x + 1$

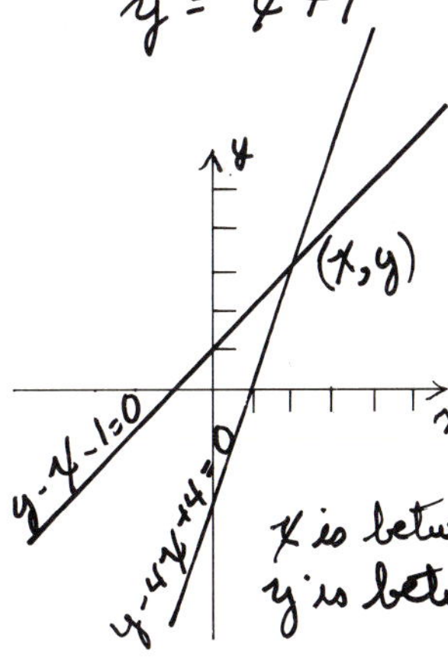

x is between 1 and 2
y is between 2 and 3

Additional Exercises

Determine the simultaneous solution (if it exists) of each of the following pairs of equations by graphing.

1. $y = x + 1$
   $y = 2x - 1$
2. $y = x + 4$
   $y = 3x - 2$
3. $y = x + 1$
   $y = x + 2$
4. $y = 2x - 3$
   $y = -x$
5. $y = 5$
   $y = x$
6. $y = -2x + 3$
   $y = x$
7. $y = 3x$
   $y = -x + 4$
8. $y = 2x - 3$
   $y = -2x + 1$
9. $x + y = 20$
   $x - y = -4$
10. $2x + y - 7 = 0$
    $x - 2y + 4 = 0$

Lesson 8

Simultaneous Solutions of Linear Equations by Algebra

In Lesson 7 we found the simultaneous solution of two equations by graphing. In this lesson we are going to find the simultaneous solution of two equations without graphing. We will find the solution algebraically.

EXAMPLE 1 Find the simultaneous solution of the following two equations algebraically: $y = x + 1$ and $y = -x + 3$

If there is a point (x, y) that makes both equations true, then for that point the y's on the left side both stand for the same number. So the right sides are equal to each other. So we write

$$x + 1 = -x + 3$$

When we solve this equation, we will get the x-coordinate of the simultaneous solution.

$$\begin{array}{lrcr} & x + 1 & = & -x + 3 \\ \text{Add } +x & +x & & +x \\ \hline & 2x + 1 & = & +3 \\ \text{Add } -1 & -1 & & -1 \\ \hline & 2x & = & +2 \end{array}$$

Divide by 2.

$$\frac{2x}{2} = \frac{+2}{2}$$

$$x = 1$$

So the x-coordinate of the simultaneous solution of the two equations is 1. Now, to find the y-coordinate of the solution we go back to either one of the two equations, say, the first one, $y = x + 1$. Since we know x is 1, we put 1 in for x.

$$y = x + 1$$

Put 1 in for x.

$$y = 1 + 1$$
$$y = 2$$

So the y-coordinate is 2, and the simultaneous solution of the two equations is (1, 2). To be sure that there are no mistakes, it is a good idea to check that (1, 2) really makes both equations true. Put 1 in for x and 2 in for y.

$y = -x + 3$		$y = x + 1$	
$2 = -1 + 3$		$2 = 1 + 1$	
$2 = 2$	True	$2 = 2$	True

The simultaneous solution (1, 2) checks. Compare this with Example 1 of Lesson 7.

EXAMPLE 2 Find the simultaneous solution of the following equations algebraically: $y = 3x$ and $y = -2x - 5$

If there is a point that makes both of these equations true, then for that point $3x = -2x - 5$ must also be true. To find what makes it true, we solve the equation.

$$\begin{array}{lrcr} & 3x & = & -2x - 5 \\ \text{Add } 2x & +2x & & +2x \\ \hline & 5x & = & -5 \end{array}$$

Divide by 5.

$$\frac{5x}{5} = \frac{-5}{5}$$

$$x = -1$$

So the x-coordinate of the simultaneous solution is -1. To find the y-coordinate we go back to either one of the original equations and put in -1 for x. We will use the first equation because it is simpler.

Since $\quad x = -1$

then $\quad y = 3x$

becomes $y = 3(-1)$

$\quad y = -3$

So the y-coordinate of the solution is -3, and the simultaneous solution of the equation is $(-1, -3)$. To check we put in -1 for x and -3 for y.

$$\begin{array}{ll} y = 3x & y = -2x - 5 \\ -3 = 3(-1) & -3 = -2(-1) - 5 \\ -3 = -3 \quad \underline{\text{True}} & -3 = 2 - 5 \\ & -3 = -3 \quad \underline{\text{True}} \end{array}$$

The simultaneous solution is $(-1, -3)$. See Example 2 of Lesson 7.

EXAMPLE 3 Find the simultaneous solution of the two equations $y = 3x - 4$ and $y = x + 2$.

If there is a simultaneous solution then

$$3x - 4 = x + 2$$

Add $-x$

$$\begin{array}{rcr} 3x - 4 & = & x + 2 \\ -x & & -x \\ \hline 2x - 4 & = & 2 \\ +4 & & +4 \\ \hline 2x & = & 6 \end{array}$$

Add $+4$

Divide by 2.

$$\frac{2x}{2} = \frac{6}{2}$$

$$x = 3$$

So the x-coordinate of the simultaneous solution is 3. To find the y-coordinate, we go back to one of the original equations and put in 3 for x.

$$y = x + 2$$

Put 3 in for x.

$$y = 3 + 2$$

$$y = 5$$

So the y-coordinate is 5 and the simultaneous solution of the two equations is (3, 5). To check, we put (3, 5) into both equations.

$$\begin{array}{ll} y = 3x - 4 & \\ 5 = 3(3) - 4 & \\ 5 = 9 - 4 & \\ 5 = 5 & \underline{\text{True}} \end{array} \qquad \begin{array}{ll} y = x + 2 & \\ 5 = 3 + 2 & \\ 5 = 5 & \underline{\text{True}} \end{array}$$

The simultaneous solution (3, 5) checks.

EXAMPLE 4 Find the simultaneous solution of the two equations $y = 2x + 3$ and $y = 2x - 1$.

If there is a simultaneous solution, then

$$\begin{array}{lrcl} & 2x + 3 & = & 2x - 1 \\ \text{Add } -2x & -2x & & -2x \\ \hline & 3 & = & -1 \end{array}$$

But $3 = -1$ is not true. Therefore, there is no solution. Let's look back and see what happened. Notice that the slope of each equation is 2. The two lines are parallel; they never cross, so there is no simultaneous solution. See Example 4 of Lesson 7.

EXAMPLE 5 Find the simultaneous solution of the two equations $y = -\frac{1}{3}x + 4$ and $y = 2x - 2$.

If there is a simultaneous solution then

$$-\frac{1}{3}x + 4 = 2x - 2$$

To eliminate the fraction, we multiply every term by 3.

$$3\left(-\frac{1}{3}x + 4\right) = 3(2x - 2)$$

$$3\left(-\frac{1}{3}x\right) + 3(4) = 3(2x) + 3(-2)$$

$$\begin{array}{lrcl} & -x + 12 & = & 6x - 6 \\ \text{Add } +x & +x & & +x \\ \hline & 12 & = & 7x - 6 \\ \text{Add } +6 & +6 & & +6 \\ \hline & 18 & = & 7x \end{array}$$

Divide by 7.

$$\frac{18}{7} = \frac{\cancel{7}^{1}x}{\cancel{7}_{1}}$$

$$\frac{18}{7} = x$$

So the x-coordinate of the simultaneous solution is $\frac{18}{7}$ which is $2\frac{4}{7}$. To find the y-coordinate of the solution, we go back to one of the original equations and put in $\frac{18}{7}$ for x.

$$y = 2x - 2$$

Put $\frac{18}{7}$ in for x.

$$y = 2\left(\frac{18}{7}\right) - 2$$

$$y = \frac{36}{7} - 2$$

$$y = \frac{22}{7}$$

The y-coordinate of the solution is $\frac{22}{7}$ which is $3\frac{1}{7}$. So the simultaneous solution is $\left(\frac{18}{7}, \frac{22}{7}\right)$ or $\left(2\frac{4}{7}, 3\frac{1}{7}\right)$.

Note: In Example 7, Lesson 7, we did this problem by graphing. We could not find the exact solution from the graph. We can find exact solutions by using algebra.

EXAMPLE 6 Find the simultaneous solution of the two equations $y = 0.7x + 2.3$ and $y = -1.3x + 0.3$.

If there is a simultaneous solution then

$$\begin{array}{lrcl} & 0.7x + 2.3 & = & -1.3x + 0.3 \\ \text{Add } +1.3x & +1.3x & & +1.3x \\ \hline & 2.0x + 2.3 & = & +0.3 \\ \text{Add } -2.3 & -2.3 & & -2.3 \\ \hline & 2.0x & = & -2.0 \end{array}$$

Divide by 2.0.

$$\frac{\overset{1}{\cancel{2.0}}x}{\underset{1}{\cancel{2.0}}} = \frac{-2.0}{2.0}$$

$$x = -1$$

So the x-coordinate of the simultaneous solution is -1. To find the y-coordinate of the solution, we go back to one of the original equations and put in -1 for x.

$$y = -1.3x + 0.3$$

Put -1 in for x.

$$\begin{aligned} y &= -1.3(-1) + 0.3 \\ y &= 1.3 + 0.3 \\ y &= 1.6 \end{aligned}$$

The y-coordinate of the solution is 1.6, so the simultaneous solution is $(-1, 1.6)$. To check, we put $(-1, 1.6)$ into both equations.

$$\begin{aligned} y &= 0.7x + 2.3 \\ 1.6 &= 0.7(-1) + 2.3 \\ 1.6 &= -0.7 + 2.3 \\ 1.6 &= 1.6 \quad \underline{\text{True}} \end{aligned} \qquad \begin{aligned} y &= -1.3x + 0.3 \\ 1.6 &= -1.3(-1) + 0.3 \\ 1.6 &= 1.3 + 0.3 \\ 1.6 &= 1.6 \quad \underline{\text{True}} \end{aligned}$$

The simultaneous solution of the two equations is $(-1, 1.6)$.

EXAMPLE 7 Find the simultaneous solution of the two equations $x + y = 7$ and $x - y = 3$.

To solve these, we first rewrite them in intercept-slope form.

$x + y = 7$ can be written as $y = -x + 7$
$x - y = 3$ can be written as $y = x - 3$

If $y = -x + 7$ and $y = x - 3$ have a simultaneous solution then

$$\begin{array}{lrcl} & -x + 7 &=& x - 3 \\ \text{Add } x & +x & & +x \\ \hline & +7 &=& 2x - 3 \\ \text{Add } 3 & +3 & & +3 \\ \hline & 10 &=& 2x \end{array}$$

Divide by 2.

$$\frac{10}{2} = \frac{2x}{2}$$

$$5 = x$$

So the x-coordinate is 5. Now put in 5 for x in one of the original equations.

$$\begin{array}{rcr} x + y &=& 7 \\ 5 + y &=& 7 \\ -5 \quad &=& -5 \\ \hline y &=& 2 \end{array}$$

So the x-coordinate is 5. Now put 5 in for x in one of the original equations.

$$\begin{array}{rcll} x + y &=& 7 & \\ 5 + 2 &=& 7 & \\ 7 &=& 7 & \underline{\text{True}} \end{array} \qquad \begin{array}{rcll} x - y &=& 3 & \\ 5 - 2 &=& 3 & \\ 3 &=& 3 & \underline{\text{True}} \end{array}$$

The simultaneous solution (5, 2) checks. See Example 6 of Lesson 7.

EXAMPLE 8 Find the simultaneous solution of the two equations $2y = 3x - 1$ and $y = x + 2$.

If we rewrite $2y = 3x - 1$ in the intercept-slope form, we will get an equation with fractions. Instead we multiply the other equation, $y = x + 2$, by 2.

$$y = x + 2$$

Multiply by 2.

$$2y = 2(x + 2)$$
$$2y = 2x + 4$$

Now both equations have $2y$ on the left side.

$$2y = 3x - 1$$
$$2y = 2x + 4$$

So if there is a simultaneous solution, the right sides are equal to each other.

$$\begin{array}{lrcr} & 3x - 1 &=& 2x + 4 \\ \text{Add } -2x & -2x \quad & & -2x \quad \\ \hline & x - 1 &=& 4 \\ \text{Add } +1 & +1 & & +1 \\ \hline & x \quad &=& 5 \end{array}$$

So the x-coordinate is 5. Now, put 5 in for x in one of the original equations.

$$y = x + 2$$
$$y = 5 + 2$$
$$y = 7$$

So the simultaneous solution is (5, 7). To check, we put (5, 7) into both equations.

$$\begin{aligned} 2y &= 3x - 1 \\ 2(7) &= 3(5) - 1 \\ 14 &= 15 - 1 \\ 14 &= 14 \quad \underline{\text{True}} \end{aligned} \qquad \begin{aligned} y &= x + 2 \\ 7 &= 5 + 2 \\ 7 &= 7 \quad \underline{\text{True}} \end{aligned}$$

EXAMPLE 9 The length of a rectangle is 6 meters more than the width, and the perimeter is 32 meters. What are the dimensions of the rectangle?

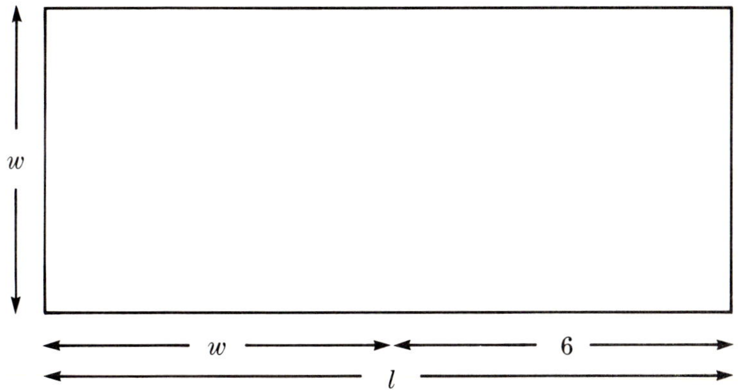

Let l stand for the length of the rectangle, and let w stand for the width. Since the length is 6 more than the width, we write

$$l = 6 + w$$

Since the perimeter is 32, we write

$$2l + 2w = 32$$

We need to find the simultaneous solution of the two equations, $l = 6 + w$ and $2l + 2w = 32$. Let's change the second equation so that we have l on the left side.

$$\begin{array}{lrcl} & 2l + 2w & = & 32 \\ \text{Add } -2w & -\ 2w & & -\ 2w \\ \hline & 2l & = & 32 - 2w \end{array}$$

Divide by 2.

$$\frac{2l}{2} = \frac{32 - 2w}{2}$$

$$l = \frac{32}{2} - \frac{2w}{2}$$

$$l = 16 - w$$

Now the two equations are $l = 6 + w$ and $l = 16 - w$.

If there is a simultaneous solution then

$$\begin{array}{lrcl} & 6 + w & = & 16 - w \\ \text{Add } +w & +\ w & & +\ w \\ \hline & 6 + 2w & = & 16 \\ \text{Add } -6 & -6 & & -6 \\ \hline & 2w & = & 10 \\ & w & = & 5 \end{array}$$

To find the value for l, we go back to one of the original equations and put in 5 for w.

$$l = 6 + w$$

becomes $l = 6 + 5$

$$l = 11$$

The simultaneous solution of the two equations is $w = 5$ and $l = 11$. So the dimensions of the rectangle are 5 meters by 11 meters.

We check: The length of the rectangle, 11 meters, is six meters more than the width, 5 meters. The perimeter is $2(11) + 2(5)$ which is $22 + 10$ or 32 meters.

Exercises

Determine the simultaneous solution (if it exists) of each of the following pairs of equations by algebraic methods.

1. $y = x - 1$
   $y = -x + 3$
2. $y = 2x + 3$
   $y = x + 1$
3. $y = \frac{1}{2}x - 2$
   $y = -2x + 3$
4. $y = 2x + 1$
   $y = -x + 4$
5. $y = 3x + 4$
   $y = -x$
6. $y = 5x + 2$
   $y = 5x - 3$
7. $y = \frac{2}{3}x - 1$
   $y = -\frac{1}{3}x - 4$
8. $y = \frac{1}{2}x + 2$
   $y = 3x - 1$
9. $y = 0.5x + 1.5$
   $y = -0.5x + 4.5$
10. $2x + 3y = 6$
    $x + y = 3$
11. $x + y = 6$
    $x - y = 4$
12. $y - 4x + 4 = 0$
    $y - x - 1 = 0$

Do the following word problem.

13. John buys 4 rosebushes and 3 azalea plants and pays \$24. JoAnn buys 2 rosebushes and 1 azalea plant and pays \$10. How much does each rosebush and each azalea plant cost?

Answers to Exercises

(1.) $y = -x + 3 \qquad y = x - 1$

$$-x + 3 = x - 1$$
$$+x \qquad +x$$
$$3 = 2x - 1$$
$$+1 \qquad +1$$
$$4 = 2x$$
$$\frac{4}{2} = \frac{2x}{2}$$
$$2 = x$$
$$x = 2$$

$$y = -x + 3$$
$$y = -(2) + 3$$
$$y = -2 + 3$$
$$y = +1 \qquad (2, 1)$$

(2.) $y = 2x + 3 \qquad y = x + 1$

$$2x + 3 = x + 1$$
$$-x \qquad -x$$
$$x + 3 = +1$$
$$-3 \qquad -3$$
$$x = -2$$

$$y = 2x + 3$$
$$y = 2(-2) + 3$$
$$y = -4 + 3$$
$$y = -1 \qquad (-2, -1)$$

(3.) $y = \frac{1}{2}x - 2 \qquad y = -2x + 3$

$$\tfrac{1}{2}x - 2 = -2x + 3$$
$$+2x \qquad +2x$$
$$2\tfrac{1}{2}x - 2 = +3$$
$$+2 \qquad +2$$
$$2\tfrac{1}{2}x = 5$$
$$x = 2$$

$$y = -2x + 3$$
$$y = -2(2) + 3$$
$$y = -4 + 3$$
$$y = -1 \qquad (2, -1)$$

(4.) $y = 2x + 1 \qquad y = -x + 4$

$$2x + 1 = -x + 4$$
$$+x \qquad +x$$
$$3x + 1 = +4$$
$$-1 \qquad -1$$
$$3x = 3$$
$$\frac{3x}{3} = \frac{3}{3}$$
$$x = 1$$

$$y = 2x + 1$$
$$y = 2(1) + 1$$
$$y = 2 + 1$$
$$y = 3 \qquad (1, 3)$$

(5.) $y = 3x + 4 \qquad y = -x$

$$3x + 4 = -x$$
$$\underline{\quad -4 \qquad\quad -4}$$
$$3x = -x - 4$$
$$\underline{+x \qquad +x}$$
$$4x = -4$$
$$x = -1$$

$$y = 3x + 4$$
$$y = 3(-1) + 4$$
$$y = -3 + 4$$
$$y = +1 \qquad (-1, +1)$$

(6.) $y = 5x + 2 \qquad y = 5x - 3$

$$5x + 2 = 5x - 3$$
$$\underline{-5x \qquad -5x}$$
$$+2 = -3$$

false: no solution

These two lines do not meet!

(7.) $y = \frac{2}{3}x - 1 \qquad y = -\frac{1}{3} - 4$

$$\frac{2}{3}x - 1 = -\frac{1}{3}x - 4$$
$$\underline{+\frac{1}{3}x \qquad +\frac{1}{3}x}$$
$$1x - 1 = -4$$
$$\underline{+1 \qquad +1}$$
$$x = -3$$

$$y = \frac{2}{3}x - 1$$
$$y = \frac{2}{3}(-3) - 1$$
$$y = -2 - 1$$
$$y = -3 \qquad (-3, -3)$$

(8.) $y = \frac{1}{2}x + 2 \quad y = 3x - 1$

$\frac{1}{2}x + 2 = 3x - 1$

$2(\frac{1}{2}x + 2) = 2(3x - 1)$

$2(\frac{1}{2}x) + 2(2) = 2(3x) + 2(-1)$

$x + 4 = 6x - 2$

$-x \qquad -x$

$4 = 5x - 2$

$+2 \qquad +2$

$6 = 5x$

$\frac{6}{5} = \frac{5x}{5}$

$\frac{6}{5} = x$

$y = 3(\frac{6}{5}) - 1$

$y = \frac{18}{5} - 1$

$y = \frac{13}{5} \quad (\frac{6}{5}, \frac{13}{5})$

(9.) $y = 0.5x + 1.5 \quad y = -0.5x + 4.5$

$0.5x + 1.5 = -0.5x + 4.5$

$+0.5x \qquad +0.5x$

$1.0x + 1.5 = 4.5$

$-1.5 \qquad -1.5$

$1.0x = 3.0$

$x = 3$

$y = 0.5(3) + 1.5$

$y = 1.5 + 1.5$

$y = 3 \quad (3, 3)$

(10.) $2x + 3y = 6 \quad x + y = 3$

$3y = 6 - 2x \quad y = 3 - x$

$3y = 9 - 3x$

$6 - 2x = 9 - 3x$

$+2x \qquad +2x$

$6 = 9 - x$

$-9 \qquad -9$

$-3 = -x$

$x = 3$

$2x + 3y = 6$

$3y = 6 - 2x$

$3y = 6 - 2(3)$

$3y = 6 - 6$

$3y = 0$

$y = 0 \quad (3, 0)$

(11) $x + y = 6 \quad x - y = 4$

$y = 6 - x \quad y = x - 4$

$$6 - x = x - 4$$
$$+x \quad +x$$
$$6 = 2x - 4$$
$$+4 \quad +4$$
$$10 = 2x$$

$$\frac{10}{2} = \frac{2x}{2}$$
$$5 = x$$
$$x = 5$$

$$x + y = 6$$
$$5 + y = 6$$
$$y = 6 - 5$$
$$y = 1 \quad (5, 1)$$

(12.) $y - 4x + 4 = 0 \quad y - x - 1 = 0$

$y = 4x - 4 \quad y = x + 1$

$$4x - 4 = x + 1$$
$$-x \quad -x$$
$$3x - 4 = 1$$
$$+4 \quad +4$$
$$3x = 5$$

$$\frac{3x}{3} = \frac{5}{3}$$
$$x = \frac{5}{3}$$
$$y = x + 1$$
$$y = \frac{5}{3} + 1$$
$$y = \frac{8}{3} \quad \left(\frac{5}{3}, \frac{8}{3}\right)$$

(13.) Let r be the cost of one rosebush and a be the cost of one azalea plant. Then,

$$4r + 3a = 24 \text{ and } 2r + 1a = 10$$

Let's get a on the left side of each equation

$$\begin{array}{rcl} 4r + 3a & = & 24 \\ -4r & & -4r \\ \hline +3a & = & 24 - 4r \end{array} \quad \text{and} \quad \begin{array}{rcl} 2r + 1a & = & 10 \\ -2r & & -2r \\ \hline a & = & 10 - 2r \end{array}$$

If we multiply the second equation by 3, we get

$$3a = 24 - 4r \quad \text{and} \quad 3a = 30 - 6r$$

so

$$\begin{array}{rcl} 24 - 4r & = & 30 - 6r \\ -24 & & -24 \\ \hline -4r & = & 6 - 6r \\ +6r & & +6r \\ \hline 2r & = & 6 \\ r & = & 3 \end{array}$$

put $r = 3$ into $2r + 1a = 10$ and get

$$\begin{aligned} 2(3) + a &= 10 \\ 6 + a &= 10 \\ a &= 4 \end{aligned}$$

so the rosebush costs \$3 and the azalea plant costs \$4.

Additional Exercises

Determine the simultaneous solution (if it exists) of each of the following pairs of equations by algebraic methods.

1. $y = 2x + 1$
   $y = x + 3$
2. $y = 3x - 2$
   $y = x + 2$
3. $y = 4x$
   $y = 3x - 1$
4. $y = 5x - 4$
   $y = \frac{1}{2}x + 5$
5. $y = -x$
   $y = 4x + 5$
6. $y = 3x + 4$
   $y = 3x - 2$
7. $y = \frac{3}{4}x - 2$
   $y = \frac{-1}{4}x + 2$
8. $y = \frac{1}{4}x + 1$
   $y = 2x - 3$
9. $y = 0.6x - 3.1$
   $y = -1.4x + 0.9$
10. $x + y = 4$
    $x - y = 2$
11. $2x + y = 6$
    $3x - y = 4$
12. $y - 3x + 3 = 0$
    $y - 2x + 2 = 0$

Do the following word problem.

13. The difference between two numbers is 8. If the smaller number is increased by four times the larger number, the result is 2. Find the numbers.

Index